# THE
# ROVING MIND

# ISAAC ASIMOV

# THE ROVING MIND

### Preface by Paul Kurtz

With Tributes by
Carl Sagan, Stephen Jay Gould, Arthur C. Clarke,
Frederik Pohl, Martin Gardner,
L. Sprague de Camp, Kendrick Frazier,
Harlan Ellison, James Randi,
Donald Goldsmith, and E. C. Krupp

## NEW EDITION

Prometheus Books

59 John Glenn Drive
Amherst, New York  14228-2197

Published 1997 by Prometheus Books

01 00 99 98 97      5 4 3 2 1

Library of Congress Cataloging-in-Publication Data

Asimov, Isaac, 1920–1992
    The roving mind / Isaac Asimov. — Rev. ed.
        p.    cm.
    ISBN 1-57392–181–5 (pbk. : alk. paper)
    1. Science—Miscellanea. I. Title.
Q173.A78     1997
081—dc21                          97–41506
                                         CIP

To the good people of the
Committee for the Scientific Investigation of
    Claims of the Paranormal,
an island of sanity in a sea of nonsense

# Contents

**Part VI   The Future**

**Part VII  Personal**

# Foreword

## Isaac Asimov: Science Popularizer, Skeptic, and Rationalist

Isaac Asimov was unique in America and the world. He was the preeminent popularizer of science in the twentieth century, having authored thousands of articles and approximately five hundred books. Asimov was born on January 2, 1920, in Petrovich, Russia, and brought to the United States by his parents when he was three years old. He lived in Brooklyn, New York, took a B.A. from Columbia University in 1939, and a Ph.D. in 1948. He became a professor of biochemistry at Boston University, though as his writing career developed, he retained his professorship without any teaching duties.

He was best known at first for his science fiction. His first stories were published in 1939, and his first book, *Pebble in the Sky,* not until 1950. His famous trilogy, *Foundation, Foundation and Empire,* and *Second Foundation,* was published in 1952–1953.

A steady stream of books flowed from his pen on a wide range of topics. His popular interpretations of science were especially impressive. He had a prodigious memory. The first drafts of his writings became largely his final drafts, with only a few minor corrections.

Asimov's role as a popularizer and a proponent of science should not be underestimated nor denigrated. Many scientists involved in their own specialties are loath to be known as popularizers, worrying about the barbs of their professional colleagues. Criticisms did not bother Asimov, who willingly assumed the role of educating the public in the methods and outlook of science. In the nineteenth century T. H. Huxley played a similar role in England, especially in defending Darwin and evolution. And in the twentieth century

Carl Sagan, Richard Dawkins, and Stephen Jay Gould (all Fellows of the Committee for the Scientific Investigation of Claims of the Paranormal [CSICOP]) have attempted to fulfill the same role. The public understanding and support of science is essential if scientific research is to continue. This is particularly true in a democracy, which depends on an educated citizenry to make wise choices. Unfortunately, large sectors of the public are scientifically illiterate—all the more reason why the popularization and interpretation of science is vital.

Although Asimov was one of the leading science-fiction writers of the twentieth century, he was careful to distinguish science fiction from concrete reality. He and I did a joint radio call-in interview in New York City several years ago. I was amused by his comment that, although he spun out his tales of science fiction, which sometimes included ESP, telepathy, and alien encounters, he "never imagined people would *believe* that crap." Regretfully, today TV broadcasts and quasi-"documentaries" do not make a distinction, and all too many people accept science fiction that is uncorroborated by evidence as true.

Asimov made a careful distinction between an open mind, receptive to new ideas and the speculative imagination, on the one hand, and confirmed hypotheses, tested by rigorous methods and found to be logically coherent, on the other. An important essay in this volume, "The Role of the Heretic," defends the importance of heresy within science. But Asimov makes a vital distinction between "endoheretics" and "exoheretics." Endoheretics emerge within the professional world of science and are subject to punishment by the received orthodoxy. But they eventually win out because their heretical views are tested by experimental replication and peer review—Galileo and Darwin are notable examples of this process. Exoheretics, on the other hand, stand outside of science and attack its orthodoxy, but their theories are never accepted by their peers because they are unable to stand the gauntlet of testing. This applies, for example, to Velikovsky.

I should say something about Asimov's skepticism concerning religion. Isaac Asimov was a strong atheist. He gladly endorsed the "Secular Humanist Declaration," which I drafted in 1980, and he was a contributor to *Free Inquiry* magazine, published by the Council for Secular Humanism. Later he was elected a Humanist Laureate of the International Academy of Humanism. When I visited him in his apartment to interview him for an article in 1982, he said that, although all too many skeptical atheists stay in the closet because they think their views are not socially respectable, he was going to express his own religious skepticism forcefully. Thus, he made clear that he did not believe in God or immortality of the soul, and that he thought that the Bible was full of contradictions and factual errors. He was correct. Many skeptics apply their skeptical doubt to only limited areas, but they are fearful of offending the powers that be in sensitive areas such as religion.

Asimov died on April 6, 1992. The important role that he played in interpreting science and defending skepticism is keenly felt today. For the present period is one in which science and technology are increasingly under attack by the purveyors of antiscientific attitudes, and there is a vast confusion in the public mind between genuine science and pseudoscience. Moreover, today the borderlines between science fiction and reality are increasingly blurred by a media onslaught in which pure nonsense parading as science undermines critical thinking and scientific rationality.

It is for this reason that Prometheus Books has decided to reissue *The Roving Mind,* first published in 1983. In this remarkable collection of sixty-two essays Asimov allows his imagination to roam freely, demonstrating his inquisitive and creative mind. In them he discusses a wide range of topics, such as creationism and the assault on evolutionism in the schools, censorship, extraterrestrial life and UFOs, technophobia, and antiscience. There are also perceptive essays on understanding the cosmos, Pluto, Jupiter, relativity theory, black holes, and hyperspace, as well as essays on futurism and a defense of cloning. This collection concludes with reminiscences of his personal life.

Included in this volume for the first time are several tributes by outstanding writers, many of whom knew him personally. These first appeared in the *Skeptical Inquirer* and they provide insightful evaluations of his life and work. It is clear that Isaac Asimov affected all those who knew him. This was not only because of his impressive writing virtuosity, but because of the incisive mind and brilliant wit that he displayed in personal encounters.

My own first contact with Isaac was about twenty-five years ago when I asked him to join the humanist movement. Later, when I founded CSICOP in 1976, I asked him if he would join our ranks, and his response was immediate and affirmative. CSICOP became the leading critic of paranormal and pseudoscientific claims. Asimov remained a strong supporter of the *Skeptical Inquirer,* published by CSICOP, and invariably responded generously to our financial appeals. Indeed, he dedicated the original edition of this book to the Committee. I only wish that Isaac had been able to participate in the national CSICOP conferences; but as is well known by his friends, he was fearful of flying. We were planning to build a CSICOP conference around him in New York City, and regret that he died before we could.

I am pleased that Prometheus Books was able to publish five books by Asimov. In addition to *The Roving Mind,* we published *The Tyrannosaurus Prescription*; *Past, Present, and Future*; *Tales of the Occult*; and *Election Day 2088.* All the manuscripts arrived in near-perfect form, needing little editing. He is remembered not only as a master of the English language, but a dedicated defender of reason, science, and skepticism.

Paul Kurtz
Publisher

# —— A Celebration of Isaac Asimov ——
## A Man for the Universe

**Kendrick Frazier**

Isaac Asimov was the master science educator of our time, and perhaps of all time.

Fame came to him early for his science fiction. To me his *I, Robot* (a collection of related stories), not the more renowned *Foundation* series, was his most memorable fiction, just ahead of such works as *The Martian Way, The Stars, Like Dust,* and *The Gods Themselves.*

But it was his science fact, particularly his science essays, that taught millions of people science. They turned me on to science as no science teacher ever did. In my freshman year in college I was a physics major, but I suffered a rude shock: the professors didn't make physics clear and interesting the way Asimov did. Suddenly it was a confusing hodgepodge of formulas and complex terms—not the orderly historical progression of people and related concepts that all science was with Asimov. Asimov had spoiled me! I hadn't expected this. For my interests, needs, and tastes, Asimov's approach was better, and to this day I still think the historical, cultural approach to teaching science has the most merit for many kinds of students.

These tributes to Isaac Asimov were originally published in *Skeptical Inquirer* 17, no. 1 (Fall 1992): 30–45, shortly after Asimov's death on April 6, 1992. Reprinted with permission of the *Skeptical Inquirer* and the following: Arthur C. Clarke; Harlan Ellison; the estate of Carl Sagan; and James Randi, President, James Randi Educational Foundation, Fort Lauderdale, Florida (http://www.randi.org).

Anyway I soon found myself a fledgling science writer rather than a would-be scientist. I took some comfort from the fact that Asimov, believing he'd probably not make a first-rate laboratory chemist, had taken the same path, except that he had a Ph.D. in chemistry. Sometime in those formative college years, about 1961, I wrote him. I asked if I should get a degree in science first before going into science journalism. I was astonished and joyous when I immediately got a card back signed "Isaac Asimov," saying yes, I should. I didn't take his advice.

But I did continue to read and learn from him. *The Intelligent Man's Guide to Science,* first published in 1960, was his systematic effort to cover all fields of the physical and biological sciences in one readable volume. It was exceedingly popular and became the model for writing about science for the "intelligent layman." Here his phenomenal breadth of knowledge, easy grasp of complex subject matter, and ability and determination to write directly, clearly, and simply for the nonscientist shone like a brilliant beacon. In three revised editions over nearly three decades, the book kept up with the rapid advance of science. Its title also evolved, regrettably losing the direct appeal to lay intelligence but thankfully dropping the unconscious sexist bias, and the current edition (all 940 pages) is titled simply *Asimov's New Guide to Science.*

His 941-page *Asimov's Biographical Encyclopedia of Science and Technology,* now in a second revised edition, sketches the lives and achievements of 1,510 great scientists from antiquity to modern times. It's not just highly readable—that's an Asimov trademark—but he arranged it chronologically, not alphabetically nor by subject, so that a careful reader can get a sense of the historical flow of ideas. Asimov believed that telling the history of science through the scientific contributions of the people who made it "stress[es] the fact that scientific knowledge is the painfully gathered product of thousands of wonderful, but fallible, human minds."

My favorite Asimov writings were his monthly science essays published in the *Magazine of Fantasy and Science Fiction.* They have appeared for thirty-three years. Only occasionally did I find the magazine itself, but about every eighteen months a new collection of the essays was published as a Doubleday book, and I eagerly awaited each new one. The twenty-fifth book in this series, *Out of the Everywhere,* was published in April [1992] as a Pinnacle paperback. Just a few of the other ones: *Fact and Fancy* (the first), *The Left Hand of The Electron, The Stars in Their Courses, The Planet That Wasn't, Quasar, Quasar Burning Bright, X Stands for Unknown, Far As Human Eye Could See,* and *The Relativity of Wrong.* (The last book's title essay was published in the Fall 1989 *Skeptical Inquirer,* and his "Asimov's Corollary," about fringe-science, was published in our Spring 1979 issue; Isaac always cheerfully granted me permission to reprint any of these essays.)

To this day I'd recommend any of these books as one of the best ways for someone interested in science to begin to learn about it in some historical depth. His 1979 F&SF anthology, *The Road to Infinity,* contained his annotated listing of the first 244 of these essays.

I loved the way he started each essay with a personal anecdote. Here his wit and humor, his jokes about his ego and intellect, the esteem he accorded a desire to learn, his masterful put-downs of those who *willfully* demonstrated ignorance about science, his cheerful embrace of the values of reason and rationality, all came through in entertaining style. It was the "Good Doctor" in a quiet conversation with the "Gentle Reader." Then he'd cleverly segue to the subject matter at hand, whatever it might be, always beginning at the beginning, with a historical approach. He was an innate storyteller, and a very orderly one. Unlike textbooks, here too he told of the people who did science and the way the concepts developed and built upon one another over time. What a wonderful way to teach! And to be taught!

These essays were more than just expositions. Like good science fiction, they were filled with provocative ideas. An example is the title essay in *The Tragedy of the Moon.* The tragedy Asimov refers to is that early people, by seeing that the moon goes around the earth (as the sun, planets, and stars also appear to), were led by their senses to believe that we are the center of the universe, an anthropocentrism whose effects remain to this day. What if, he asked, Venus had had a moon of the same relative size? Such a moon could have been visible to the eye from the earth, and people would have had a clear example of another heavenly body besides the earth having something revolving around it. The history of human thought and culture might have been noticeably different. Balancing this "tragedy of the moon" is "The Triumph of the Moon." In this companion essay, Asimov considered how life itself may owe a seminal debt to the moon; shallow tide pools, whose ebbs and flows are caused mainly by lunar tides, may have served as the place of molecular self-assembly that resulted in the first life-forms on earth. Perhaps a large moon is necessary for life on a planet to take hold.

Asimov did not like to travel, and seldom did so, preferring instead to roam about the universe in his imagination, and not coincidentally to keep at the typewriter (and only much later the word processor) day and night. Nevertheless, he was an outgoing, ebullient man, with a razor wit and world-class sense of humor (yes, he wrote books of limericks and books about humor). The first time I ever saw him was at a science-fiction convention in Washington. There he and Harlan Ellison were going at it from opposite ends of a giant, standing-room-only ballroom in what I can only describe as an insult-hurling contest. The game was to see who could get the best of the other with the wittiest and most penetrating barbs. Harlan, who is very, very good at this, had here met his match. It was all great fun.

It was always amusing to witness the banter between Asimov and his good friend Arthur C. Clarke. They constantly teased each other in print over who was the better science-fiction writer and the better science-fact writer. Finally, they came to an equitable agreement. It was known as the Clarke-Asimov Treaty. As a result, Clarke's nonfiction book *Report on Planet Three* contained this dedication: "In accordance with the terms of the Clarke-Asimov Treaty, the second-best science writer dedicates this book to the second-best science-fiction writer."

Clarke introduced Asimov at a conference in 1974 as a four-typewriter threat, "the only man who can type separate books simultaneously with his two feet as well as his two hands." He calculated that Asimov to that point had been responsible for deforestation amounting to "5.7 times ten to the sixteenth microhectares. . . . All those beautiful trees, turned into Asimov books." Asimov responded that Clarke's introduction was the very worst kind—long and clever—and intentionally so. And he told his audience Clarke was the kind of man who, upon receiving a seventy-five-page crank letter in an indecipherable handwritten scrawl on onion-skin paper purporting to explain the entire universe, would reply saying he couldn't give the theory the attention it deserved "but my friend Isaac Asimov is interested in just this sort of thing" and give Asimov's address. (This whole amusing exchange appears in Clarke's *The View from Serendip,* and my thanks to Arthur Clarke for reminding me of it.)

Clarke also thought our readers might enjoy this limerick that Asimov wrote on a paper napkin and gave to him (he still has it) at a science-fiction publisher's dinner in New York in 1977:

> Old Arthur C. Clarke of Sri Lanka
> Now sits in the sun sipping Sanka
> Enjoying his ease
> Excepting when he's
> Receiving pleased notes from his banker.

When CSICOP was founded by Paul Kurtz in 1976, Asimov was one of the original founding Fellows. Even with such luminaries as Martin Gardner, James Randi, Carl Sagan, and B. F. Skinner as Fellows as well, he was perhaps the most famous. I came on board the next year as editor and soon thereafter wrote Asimov asking him to be a *Skeptical Inquirer* consulting editor. He readily agreed. The next year I met him for the first time when he dropped in on a meeting of the CSICOP Executive Council in a midtown Manhattan hotel. For a man thought to have such a formidable ego, his first words to me were unexpected: "Oh, you're my editor!" This was a joke, of course, for no writer ever less needed an editor, much less me, but I have never forgotten that gentle kindness.

Asimov mostly listened that morning (another attribute one would have been led to believe was uncharacteristic). In chitchat afterward about creationism, he quickly caught a mortal flaw in a typical creationist argument against evolution. "The earth isn't a closed system!" he exclaimed with an exasperated laugh. "The sun provides energy from the outside. Nothing about evolution violates the laws of thermodynamics."

In 1983 Asimov put together a collection of essays for Prometheus Books titled *The Roving Mind.* The essays covered a wide variety of topics that included fringe science, technology, the future, and social issues involving science. His dedication read:

> To the good people of the
> Committee for the Scientific
> Investigation of Claims of the Paranormal,
> an island of sanity in a sea of nonsense

One of my disappointments is that CSICOP never held one of its big annual conferences in New York City so that we might have had Isaac as our keynote speaker—what an attraction that would have been!—and awarded him our "In Praise of Reason" Award. He richly deserved it.

When we celebrated the tenth anniversary of the *Skeptical Inquirer* in 1986, Asimov contributed an original essay. It was titled "The Perennial Fringe." While it granted the ready appeal of comforting pseudoscience ("a thumb to suck, a skirt to hold") in comparison with uncertain science, it ended with a ringing appeal that, where matters of state and democracy are concerned, we never let the forces of unreason prevail. "We must fight any attempt on the part of the fringers and irrationalists to call to their side the force of the state. . . . That we must fight to the death" (See *Skeptical Inquirer,* Spring 1986, reprinted in the *Skeptical Inquirer* anthology *The Hundredth Monkey and Other Paradigms of the Paranormal,* Prometheus Books, 1991.)

Fifteen years ago, Asimov took time out from his other works to write his autobiography. He finished it on New Year's Eve, 1977; it and a coronary had cut into his productivity somewhat: in his annual end-of-year stocktakings he noted that in 1977 he had published only ten books (!), the fewest in seven years; and in 1978, seven. The autobiography turned out to be 640,000 words long, and his editor said it would have to be published in two volumes. He playfully protested that William Shirer's *Rise and Fall of the Third Reich* was 650,000 words long, "according to my careful word count, and that's in one volume," but to no avail. The first volume (1920–1953) became *In Memory Yet Green*; the second (1954–1978), *In Joy Still Felt.* Drawing upon daily diaries he assiduously kept and strictly chronological (that sense of order again), it's as highly readable as all his other writings and is filled with information and delightful stories from his daily life.

Asimov ended the 806th and final page of *In Joy Still Felt* with the now-poignant words, "To Be Continued, Eventually." A few pages earlier he wrote: "It is my intention, if I live to the end of the century or thereabouts, to do a third and (I suppose) final volume to be called *The Scenes from a Life.* . . . However, the vicissitudes of life are uncertain, and I may not get the chance to do that third volume. . . ." In *Joy Still Felt,* published in 1980, was Asimov book number 215. It's astonishing to realize that in his final twelve years of life he would more than match that total output. By the time of his death, early in the morning on April 6, 1992, the number of Asimov books had long since passed 460 and was quickly rising toward 500. His was a prodigious and glorious body of work, and in combination of magnitude, substance, breadth, and diversity, it is likely never to be equaled.

Isaac Asimov would countenance no illusions about the finality of death. Yet through his works and in the lives of those he affected, he will live on, forever. He, perhaps more so than any other person in history, truly is a man of, and for, the universe.

## Arthur C. Clarke

Many years ago, when introducing Isaac Asimov to a Mensa Society meeting in London, I said, "Ladies and gentlemen, there is only one Isaac Asimov." Now there is no Isaac Asimov, and the world is a much poorer place.

Isaac must have been one of the greatest educators who ever lived, with his almost half a thousand books on virtually every aspect of science and culture. His country has lost him at its moment of direst need, for he was a powerful force against the evils that seem about to overwhelm it (and much of Western society). He stood for knowledge against superstition, tolerance against bigotry, kindness against cruelty—above all, peace against war. His was one of the most effective voices against the "New Age" nitwits and fundamentalist fanatics who may now be a greater menace than the paper bear of communism ever was.

Isaac's fiction was as important as his nonfiction, because it spread the same ideas on an even wider scale. He virtually invented the science of robotics—and named it before it was born. Without preaching, he showed that knowledge was better than ignorance and that there were other defenses against violence than violence itself.

Finally, and not least, he was great fun. He will be sorely missed by thousands of friends and millions of admirers.

## Frederik Pohl

Isaac was part of my life for more than half a century. Sometimes we worked together. I was his literary agent for a while, now and then his editor. We did some writing together, too—a couple of short stories long ago, and then *Our Angry Earth* just last year—but most of my memories of Isaac are not of our professional relationship but of moments we shared. I remember huddling with him over a television set in a Boston hotel room when the first pictures of the surface of Mars were coming in, and the way he looked up at me indignantly and said, "Craters? How come neither of us thought of *craters* on Mars?" I remember a Caribbean cruise to watch the nighttime launch of Apollo 17, when I turned around just after lift-off and saw Isaac illuminated in that giant sun-burst Saturn-5 rocket flare with Bob Heinlein and Ted Sturgeon beside him; I wished I had had the intelligence to take along a camera so I could photograph those faces shining in that wonderful light And I remember the Futurian days, when all of us wanted so badly to get published. In those poverty-stricken Depression times Isaac was not only a friend, he was a valuable economic asset, because when the thirst struck and the bankroll was flat I could always walk across Prospect Park to where his parents had their candy store and get a free chocolate malted from his mother. Of course there are plenty of more sub-stantial reasons to remember Isaac—all those books, all those wonderful accomplishments—but those are some of the ones that are my own.

Isaac knew he was dying, and calmly and courageously let us know it, too. But, even though I was forewarned, when CBS woke me that Monday morning with the word that he was gone it still hurt. There has never been anyone else like him, and I don't think there ever will be again.


**Harlan Ellison**

Everything he stood for, everything he tried to teach us, prevents me from eulogizing him by way of suggesting He Has Gone to a Better Place. I'd really like to; but he won't permit it.

In the 1984 collection of his science essays, *'X' Stands for Unknown,* Isaac wrote: "There seems to be a vague notion that something omniscient and omnipotent *must* exist. If it can be shown that scientists are not all-knowing and all-powerful, then that must be the proof that something else that *is* omni-scient and omnipotent *does* exist. In other words: Since scientists can't syn-thesize sucrose, God exists.

"Well, God may exist; I won't argue the point here—"

And a year earlier, in *The Roving Mind,* he began an essay on "faith" titled "Don't You Believe?" like this:

"One of the curses of being a well-known science-fiction writer is that unsophisticated people assume you to be soft in the head. They come to you for refuge from a hard and skeptical world.

"Don't you believe in flying saucers? they ask me. Don't you believe in telepathy?—in ancient astronauts?—in the Bermuda Triangle?—in life after death?

"No, I reply. No, no, no, no, and again no."

How dare I, then, dishonor all that he was about, publicly and privately, in print and in person, for the fifty-four years that we were pals, by suggesting that at last Isaac will be able to get first-hand answers to the questions that drove him crazy throughout most of his life, from Darwin and Roentgen and Einstein and Galileo and Faraday and Tesla . . . just sitting around, shooting the breeze with the guys, as Archimedes mixes the drinks.

As it was for all of us who needed a question answered, who called Isaac at all hours of the day or night, who drowned him in requests for answers to conundrums, so it will now be for Isaac, chasing down Cervantes and Willy Shakespeare and Jesus, buttonholing them for the answers to the maybe six or seven things in the universe he didn't know. Such little fantasies might make it easier to live with his death, but it would only be balm for those of us who listened to Isaac for decades but reverted to superstition when the bullets whistled past our ears.

Gone is gone, and with the passing of Isaac, who loved us deeply enough to chivvy us toward smartness with a relentless passion, the universe has shrunk more than a little. He is gone and, as I write these words less than twelve hours later, there is no more crying left in me. Those of us who were so dear to his heart, well, we've known for many months he wouldn't be with us much longer; and we've had time to wring ourselves out. And yet there is no end to the sense of helplessness and loss.

Isaac was as much a part of the journals that decry paralogical thinking as paper and ink; and though gone, he remains with us. As he remains with the uncounted thousands of young people who read his essays and stories and went into careers of scientific inquiry, who understood the physical universe because he made it graspable, who became better able to handle their lives because he refused to allow them to accept dogma and bigotry and mendacity in place of common sense and logic.

For all of you who will mourn him in your own way, the most I have to offer is this one last anecdote of how he viewed himself and his imminent passage:

His wife, Dr. Janet Jeppson, was with him at the end, of course; and his daughter, Robyn. Janet told me, the day before he died, that toward the end Isaac had trouble speaking, could only manage a word or two from time to time. He would say *I love you* to Janet, and he would smile. But every once in a while he would murmur, "I want . . ." and never finish the sentence. "I want . . ."

And Janet would try to perceive what he needed, and she would say, "A drink of water?" or "Something to eat?" And Isaac would look dismayed, annoyed, chagrined that he couldn't put the sentence together; and after a moment he would let it slide, and forget he had spoken. Until the time came on the Sunday before he went back into the hospital for the last visit, when he managed to say, very clearly . . .

"I want . . . I want . . . Isaac Asimov."

And Janet told him he *was* Isaac Asimov, that he had *always* been Isaac

Asimov. But he looked troubled. That wasn't what he meant. Then Janet remembered that Isaac had told her, some time ago, before he began to slip into abstraction and silence, that if there ever came a time when he didn't know who he was, if there came a time when his mind was not sharp, that he wanted to be left to go to sleep quietly, that extraordinary measures should not be taken.

And Janet understood he was saying that he wanted to *be* Isaac Asimov again.

Then, in that final week before 2:30 A.M. New York time on Monday, April 6th, he was holding Janet's hand, and he looked up at her and said, very clearly, the last he would ever say, "I am Isaac Asimov."

Yes, he was. Yes, indeed, he was.

## L. Sprague de Camp

I first met Isaac Asimov on May 7, 1939, at the Queens Science Fiction League. When introduced, Isaac stood up and said: "Now you see the world's worst science-fiction writer!"

For years he made such wildly self-deprecating remarks. Willy Ley and I once chided him about it, whereupon he said: "But if I don't, people will think I'm conceited!"

Willy and I told him that he could avoid such a fate by simply not talking about himself. The advice had little visible effect, since the nineteen-year-old Isaac was an irrepressible extrovert, voluble, impulsive, and expansive.

Over the next two years, I ran into Isaac at science-fiction gatherings. On June 28, 1941, he came to Catherine's and my apartment on Riverside Drive for dinner. In his first autobiographical volume, *In Memory Yet Green,* Isaac wrote: "It was the first time I had ever been asked to visit the home of an established science-fiction writer. It was a matter of great excitement for me." Later he told someone that the reason he had such a soft spot for the de Camps was that we were the first gentiles to treat him as a social equal.

On a later dinner visit, I offered Isaac a highball. Just a little one, he said; so I poured him an ounce of rye whiskey and added a mixer. Isaac drank the dose but soon became oddly flushed and mottled. He politely took his leave but did not dare go home in what he thought was a tipsy state. He rode the subway from one end of the line to the other, making three round trips before returning home. Actually he was not intoxicated; he later learned that he had an allergy to alcohol, which kept him a virtual teetotaler all his life.

The war news was discouraging at that time. Hitler had suddenly attacked the Soviet Union along the border established when the two powers had partitioned Poland in 1939. For a month, the Nazis made huge gains and took millions of Russian prisoners. Isaac remarked that, the ways things were going, he could look forward only to an early death. Asked why, he said: "Because I'm a Jew."

Actually he was not an observant and had no supernatural beliefs; but Nazis made no such distinction.

Since Isaac became much more productive and widely read than I, the honor of that first dinner's entertainment should go to him rather than to the de Camps.

In December 1941 came Pearl Harbor. Robert Heinlein had kept in touch with an Annapolis classmate, A. B. Scoles (then a lieutenant commander), who had been appointed director of the Materials Laboratory of the Naval Air Experimental Station of the Philadelphia Naval Base. Aware of Robert Heinlein's writing career, Scoles thought: Why not get a few of these fellows with technical backgrounds, who have been writing glibly about death rays and spaceships, to go to work here and show what they can do?

So Heinlein went to work at the Materials Laboratory as a civilian engineer (the Navy refused to put him back in uniform because of his medical history), and I joined him when I finished my naval training as a lieutenant, USNR. Scoles also persuaded Isaac, then a graduate student at Columbia, to come to Philadelphia as a civilian chemist.

For three and a half years, Heinlein, Asimov, and I navigated desks and fought the war with flashing slide rules. Soon after the war, the now-defunct *Philadelphia Record* ran a feature article headed "Stranger Than Fiction." The piece derisively narrated how the Navy had hired three "mad scientists" (that is, science-fiction writers) to invent superweapons, none of which worked. There was practically not a word of truth in the article. Asimov's name was misspelled; I was wrongly identified as a University of California graduate and an "expert aerodynamicist," and so on. Asimov and I wrote angry letters, but it took a threatening call from a lawyer to make the paper backtrack.

Actually, we three were assigned to separate sections and did not work together; and there was little or no mad-scientist element in our work. I tested things like hydraulic valves for Naval aircraft, trim-tab controls, and windshield de-icers. Asimov performed the chemical jobs assigned to him. Heinlein's work was so secret that I still do not know what he did.

Our contacts thereafter were episodic: meetings at conventions; Catherine's and my occasional visits to Boston; and intermittent correspondence. In 1950, while we were living in Wallingford, Pennsylvania, Isaac visited us when he came from New York for a meeting of the American Chemical Society. I was struggling with the plot of my novel *The Glory That Was*. I appealed to Isaac, who made some sound suggestions. Since the story has been reprinted several times, including a recent new edition, the book proved a fair success, for which Isaac merits part of the credit

Years later, when Isaac had moved back to New York, I got him into the Trap Door Spiders, the all-male eating, drinking, and arguing society formed by Fletcher Pratt in 1944. Isaac remained the club's most distinguished ornament down to his death.

I considered Isaac Asimov one of my oldest, closest, and most beloved friends, although geographical separation kept us from seeing each other much more often than the monthly meetings of the Trap Door Spiders. This friendship endured despite differences of background, age, and temperament. In his youth, Isaac was noisy, brash, impulsive, and intensely emotional. As he explains in his autobiography, he could not resist the urge to show off, express opinions, make jokes, and "crack wise," even when he knew such acts to be contraproductive. I was more reserved, solitary, and introverted, although I forced myself to learn to do active things like riding and sailing. Isaac became more and more involved in his writing to the exclusion of all else. I have traveled the world; he disliked travel, avoided airplanes, and in recent decades refused to stir far from his typewriter.

My lifelong friendship with Isaac is one of my most precious memories. Of all the people I have known, I rate Isaac as the most intelligent. Added to this brilliance of mind was *character,* his utter, transparent integrity, which compelled him to do what he thought right, even at his own sacrifice. If, a century hence, someone writes about the two of us, I shall be honored to be briefly mentioned as "a friend of Isaac Asimov."

## Carl Sagan

Isaac Asimov was one of the great explainers of the age. Like T. H. Huxley, he was motivated by profoundly democratic impulses to communicate science to the public. "Science is too important," he said, paraphrasing Clemenceau, "to be left to the scientists." It will never be known how many practicing scientists today, in how many countries, owe their initial inspiration to a book, article, or short story by Isaac Asimov—nor how many ordinary citizens are sympathetic to the scientific enterprise from the same cause. For example, Marvin Minsky of MIT, one of the pioneers of artificial intelligence, was brought to his subject by Asimov's robot stories (initially conceived to illustrate human/robot partnerships and to counter the prevailing notion, going back to *Frankenstein,* of robots as necessarily malign). At a time when science fiction was mainly devoted to action and adventure, Asimov introduced puzzle-solving schemes that taught science and thinking along the way.

A number of his phrases and ideas have insinuated themselves into the culture of science—for example, his spare description of the solar system as "four planets plus debris" and his notion of one day carrying icebergs from the rings of Saturn to the arid wastelands of Mars. He wrote many science books for young people, and as editor of his own science-fiction magazine he made efforts to encourage young writers.

His output was prodigious, approaching five hundred volumes, always in his characteristic straightforward, plain-speaking syntax. Part of the reason his *Foundation* series on the decline of a galactic empire worked so well is that it was based on a close reading of Gibbon's *Decline and Fall of the Roman Empire*: A principal theme was the effort to keep science alive as the Dark Ages rolled in.

Asimov spoke out in favor of science and reason and against pseudo-science and superstition. He was not afraid to criticize the U.S. government and was deeply committed to stabilizing world population growth.

The microscopic probe he described in his novel *Fantastic Voyage*—which could enter the human bloodstream and repair tissue damage—was, sadly, not yet available at the time of his death. As someone born in grinding poverty, and with a lifelong passion to write and explain, Asimov by his own standards led a successful and happy life. In one of his last books he wrote: "My life has just about run its course and I don't really expect to live much longer." However, he went on, his love for his wife, the psychiatrist Janet Jeppson, and hers for him, sustained him. "It's been a good life, and I am satisfied with it. So please don't worry about me."

I don't. Instead, I worry about the rest of us, with no Isaac Asimov around to inspire the young to learning and to science.


### Stephen Jay Gould

My first contact with Isaac Asimov was daunting. I picked up the phone one day, and a voice bellowed: "Gould, this is Isaac Asimov. I hate you."

"Oh," I replied with astonishing lack of originality, "why so?"

"I hate you because you write so well," he said.

So I replied, "And if I had written four hundred books instead of ten, I wouldn't be paying such rapt attention to stylistic nuances either."

We both laughed and became good friends. Isaac was the best (and most copious) there has ever been—ever throughout history—in the presentation of science. Only Galileo and Huxley (maybe Medawar in our generation) matched his clarity, his verve, his dedication, and, above all, his moral sense of the rightness and power of knowledge.


### Martin Gardner

Knowing Isaac Asimov was one of my life's great benedictions. I can vividly recall our first meeting. Isaac had been reading my *Scientific American* columns, and he wanted to know what sort of formal training I had in mathematics. When I told him I had none, that I merely read what the real mathematicians were saying and then tried to dish it out in entertaining ways, he slapped his forehead. "You mean," he exclaimed, "that you are working the same racket I am?"

Isaac liked to pretend he was an egotist, but when he talked about his obviously high intelligence it was always in such amusing ways that it annoyed no one. When Isaac was about to be given an anesthetic before an operation, recalled Andy Rooney in a fine tribute to his "lovable" and "unlikely" friend, he said to the doctor, "I hope you understand this is not an average brain you're about to put to sleep." He even had a business card that said under his name, "Natural Resource." Unlike the truly conceited, Isaac never indulged in false modesty.

No modern writer has done more, or is likely to do more, to introduce people of all ages to the wonders of science and to combat the scientific illiteracy that increases every year. It has seeped into Congress. It even invaded the White House when it was occupied by the Reagans. Something is radically wrong with a nation willing to issue a stamp honoring the forgettable hound dog and drug addict Elvis Presley. Let us hope that someday our post office will have enough sense to devote a commemorative stamp to Isaac Asimov, an authentic national treasure.

## Donald Goldsmith

My acquaintance with Isaac Asimov arose from the 1974 American Association for the Advancement of Science symposium on Immanuel Velikovsky, which I helped to organize. Isaac then contributed the foreword to *Scientists Confront Velikovsky,* the book that grew out of that symposium, which I edited. During the final stages of preparing the manuscripts for publication, as I awaited Isaac's comments on my editing, he suffered a heart attack. Any other writer would have let so small a matter as this foreword—in a book edited by an unknown—quite naturally wait for his recovery, but Isaac sent me a handwritten note stating, "I have annoyed everyone by having a coronary and being committed to a hospital. Under the circumstances, I'll go along with any changes you wish made." Only Isaac Asimov could combine such generosity and organization with an ability to work on ten projects at once. We shall not see his like again.

## James Randi

I was once long ago invited to a meeting of the Trap Door Spiders, an informal group of science-fiction and mystery writers in the New York area who gathered regularly for an evening of fine food and drink, always at the home of a member. It was the custom that one visiting guest was permitted to attend, with the strict requirement that he had to, on that occasion, provide the others with a good reason for his continued existence. (The all-male Spiders held these affairs in the absence of their wives.) Apparently I was able to make an adequate case for being permitted to live on, and thus I first met Isaac Asimov, in the company of John Dickson Carr, George O. Smith, Lester Del Rey, Frederik Pohl, and other literary luminaries.

Isaac's formidable sideburns seldom stopped moving as he competed with others in improving their mutual knowledge of the world. He was, I quickly discovered, an authority on everything.

As I've always said about Isaac, he had an enormously developed ego, but he had every right to it. Although his name is now well known around the world, I was informed by my editor friend, Clayton Rawson, that there was a period in the early days when Isaac suffered the indignity of having his name incorrectly spelled. At the time that he was submitting his first stories to editors in New York, Rawson and his colleague Lester Del Rey played a joke on him by mailing him an edited version of one of his short stories with the author's name as "Asaac Isimoff." Isaac in those days had the reputation of not spending his money needlessly, but this affront brought about a rare person-to-person long-distance phone call from the alarmed author to Rawson's New York office, much to the amusement of all—except possibly the author himself.

Aware of this classic situation, I found myself one evening appearing on an early New York television program along with a number of other people, among them Isaac. Fiend that I am, I changed his name on the dressing-room list to Asaac Isimoff, and removed mine. Seated at the mirror prettying up for my appearance, I soon heard a mighty roar echoing about the stairways of the studio. "There's a damn magician on this show somewhere, and I want his heart!" bellowed the itinerant genius as he confronted this impertinence. I barely survived his wrath.

Perhaps Isaac Asimov established some sort of record for leaving his thoughts behind him, in the multitude of books, essays, stories, and manuscripts that he created. Every subject from the sun to the Bible came under his examination and was the better for it. His interaction with my life certainly served me well, and I remember him as a delightful, brilliant, and kindly man who never refused me a favor and who added to my enjoyment of, and dedication to, science.

The man never believed in survival after death or in any of the metaphysical claptrap with which he was regularly confronted by the nut fringe. Paradoxically, he had a lifelong fear of flying in an airplane yet wrote of heroes who traveled at light-speed-plus. But, for all we know, dear Isaac may now be out there among the planets and stars, characteristically chasing after some particularly interesting comet as the possible subject for his next book. If the galaxies have secrets, they may now prepare to surrender them to his scrutiny.

## E. C. Krupp

I never got to meet Isaac Asimov. I know him from his books. There is a line of twenty-two mass-market paperbacks under his byline parading on one of the shelves in my office at Griffith Observatory. (Of course, that does not count the science fiction I have at home.) That is a small fraction of his total output, but twenty-two titles by one author is a respectable showing on any shelf. I am particularly fond of *The Stars in Their Courses*. It contains classic Asimov—"Worlds in Confusion." In it, Isaac discussed the physical implications of Velikovsky's pseudoscientific ideas. Asimov's essay is a showpiece of popular science. Clarity, humor, logic, and anecdote—they are all there in spades. Some of his lines continue to lighten my heart:

> There is no belief, however foolish, that will not gather its faithful adherents who will defend it to the death.

> Gentle Reader, place all myths and legends of the human race at my disposal; give me leave to choose those which I want to use and allow me to make changes where necessary, and I will undertake to prove anything you wish proven.

> If I must choose between Immanuel Velikovsky and Cecil B. de Mille, give me de Mille, and quickly.

Thanks, Isaac. I, too, prefer de Mille.

# Introduction

I have the roving mind of the title, as well as an easy touch at the typewriter (or word-processor), and editors have found that out. The result is that although, left to myself, I would in any case deal with a wide variety of subjects, I am forced to extend myself even further by the suggestions of pleasant people who want to fill the pages of their magazines with matters they consider both important and of interest to their readers.

I might, of course, turn them down and, with an austere smile, do whatever it is I would do if I weren't being prodded. There are, unfortunately, two reasons why I do not do this.

First, I am the softest touch in the world. A little bit of flattery, a few words to the effect that only I am skilled enough to deal with the subject properly or that only I am professional enough to get it in by deadlines and I will accede to anything. And if it happens that an adequate fee is mentioned, that doesn't hurt either. (I am far too noble a soul, of course, to have the slightest regard for money, but there are numerous crass and vulgar people in the world who expect money in return for goods and services and whose bills drop softly, repeatedly, and punctually into my letterbox. It is for their sake, and not for mine, that I ever allow the subject of financial remuneration to arise.)

Second, there is that phrase about doing "whatever it is I would do if I weren't being prodded." Actually, I haven't figured out what that might be. I've considered golf, travel, lying in the sun, watching television, going to parties, and various other alternatives, and found them uniformly vile. The only thing I really want to do is to sit at a typewriter (or word-processor) and unreel my thoughts. For that reason, I am actually very grateful to editors who prod me and make it easier for me to do so.

1

It is one of my many amiable literary characteristics that I gather various related articles together, now and then, and persuade a publisher to put out a book of my collected essays. This is a rather old-fashioned thing to do these days and few such books are published, on the whole. Such are my powers of persuasion, however, that I have, as of now, managed to get twenty-eight essay collections into print. This may make me as the most prolific essayist in history.

You may easily believe that I am carrying the matter entirely too far, and I agree with you—but it isn't my fault. As I told you, it's entirely the fault of the editors. Despite twenty-eight collections that contain nearly five hundred different essays, I find, to my horrified discomfiture, that under editorial pressure the darn things accumulate faster than I can push them into collections.

Which brings us to the present—

Victor Gulotta, of Prometheus Books, sent me a blandishing letter suggesting that I might want to put together a collection of essays for him.

Of course! The very thing!

I rounded up sixty-two essays at once, almost all of which were published within the last half-dozen years, and none of which have been included in any previous collection. A few of them, in fact, have not been published at all because, although they were asked for by editors, what I turned out did not quite fit their editorial policy—either because I'm not perfect or because their editorial policy is dead wrong (undoubtedly the latter in every case).

The sixty-two essays are, of course, a wild miscellany, ranging from the polemical to the persuasive, from the speculative to the realistic.

There is, I must admit, some overlapping. Since I have written the essays for a wide variety of periodicals and have addressed a wide variety of audiences, I am tempted to repeat my favorite viewpoints in different ways, and some of the repetition will be laid out before you. I may even contradict myself on rare occasions.

Please forgive me these flaws, which are, after all, inherent in this sort of book.

If it will do you any good, please remember that you are perfectly free to send me an eloquent letter denouncing any views with which you are eccentric enough to disagree, refuting any scientific points you may think you find in error, and sneering at any stylistic excesses or deficiencies you unaccountably decide to consider insupportable. I don't say that I will enjoy such letters, but I will read them, and, if time permits, I may even respond.

Now it is time for me to let you go and to send you on into the body of the book—but not without a warning.

If, having read this introduction, you have decided I am genial, good-natured, and lovable (and you are right), let me warn you that the first article you will read is nothing of the sort.

So take a deep breath before you go on—

# Part I

# The Religious Radicals

# 1

# The Army of the Night

Scientists thought it was settled.

The universe, they had decided, is about fifteen billion years old and the earth itself is nearly five billion years old. Simple forms of life came into being over three billion years ago, having formed spontaneously from non-living matter. They grew more complex through slow evolutionary processes and the first hominid ancestors of humanity appeared over four million years ago. *Homo sapiens* itself, the present human species, people like you and me, have walked the earth for at least 50,000 years.

But it isn't settled. There are Americans who believe that the earth is 10,000 years old at most; that human beings and all other species were brought into existence by a divine Creator as eternally separate varieties of beings; that there has been no evolutionary process and there never was. They are creationists, and they call themselves "scientific" creationists.

Such creationists are a growing power in the land and are demanding that schools be forced to teach their views. State legislatures, mindful of votes, are showing signs of caving in before them. In Arkansas, in Iowa, in Florida, in California, strong movements are on the way to legislate the teaching of creationism.

Is this really something to fear? Surely only a small minority of the nation is creationist — not vanishingly small, however. Jerry Falwell's television pulpit alone is supposed to have fifteen million viewers, and in parts of the so-called Bible Belt creationists are in the majority.

They make up a fervid and dedicated group of followers, convinced beyond argument of both their rightness and righteousness, and able to use their simplistic conservatism and sloganistic patriotism to lure to their side allies who are not directly interested in creationist views. Societies have

5

been disrupted and taken over by smaller groups than this when the majority has been apathetic and falsely secure.

To those who are trained in science, creationism seems a bad dream, a sudden coming back to life of a nightmare, a renewed march of an Army of the Night risen to challenge free thought and enlightenment.

The scientific evidence for the age of the earth and for the evolutionary development of life seems overwhelming to scientists. How can anyone question it? What are the arguments the creationists use? What is the "science" that makes their views "scientific"? Here are some of them.

1. *The argument from analogy.* A watch implies a watchmaker, say the creationists. If you were to find a beautifully intricate watch in the desert, far from habitation, you would be sure that it had been fashioned by human hands and somehow left there. It would pass the bounds of credibility that it had simply formed, spontaneously, from the sands of the desert.

By analogy, then, if you consider humanity, life, earth, and the universe, all infinitely more intricate than a watch, you can far less believe that it "just happened." It, too, like the watch, must have been fashioned, but by more-than-human hands; in short by a Divine Creator.

This argument seems unanswerable and it has been used (even though not often explicitly expressed) ever since the dawn of consciousness in order to fashion a world of gods and demons.

Thus — To sprinkle water on flowers requires a watering-can; therefore, the rain descends from a divine watering-can held by a god and can be yielded or withheld at divine whim.

To cool your porridge with a breath requires human lungs; therefore the wind is the product of the divine lungs of a god.

To travel long distances at a good clip requires a horse and carriage with yourself at the reins; therefore the sun in crossing the sky requires a flaming horse and carriage with a god at the reins.

One can go on and on. To have explained to prescientific human beings that the wind and the rain and the sun follow the laws of nature and do so blindly and without a guiding mind would have been utterly unconvincing to them. In fact, it might well have gotten you stoned to death as a blasphemer.

This argument reduces God to a one-syllable sound meaning "I don't know."

There are many aspects of the universe that still can't be explained satisfactorily by science; but ignorance implies only ignorance that may some day be conquered. To surrender to ignorance and call it God has always been premature up to this time, and it remains premature today.

In short, the complexity of the universe and one's inability to explain it in full, is not, in itself, an argument for a Creator.

2. *The argument from general consent.* Some creationists point out that belief in a Creator is general among all peoples and all cultures. Surely this

unanimous craving hints at a great Truth. There would be no unanimous belief in a lie.

General belief, however, is not really surprising. From the analogy argument previously mentioned, any people, any group, that considers the existence of the world would assume it to have been created by a god or gods, just as human beings themselves fashion hunting spears and pottery.

Naturally, each group invents full detail for the story and no two creation tales are alike. The Greeks, the Norsemen, the Japanese, the Hindus, the American Indians, and so on and so on and so on, all have their own creation myths, and all of these are recognized by Americans of Judeo-Christian heritage as "just myths."

The ancient Hebrews also had a creation tale—two of them, in fact. There is a primitive Adam-and-Eve-in-Paradise story, with man created first, then animals, then woman. There is also a poetic tale of God fashioning the universe in six days, with animals preceding man, and man and woman created together.

These Hebrew myths are not inherently more credible than any of the others, but they are *our* myths and the only ones that the creationists are interested in or (in most cases) have heard of, and the only ones they want to propagate.

Surely, if it is general consent that proves the existence of a Creator, then general dissent disproves every other aspect of creation, since no culture believes any creation myth but its own.

In fact, if you come right down to it, general consent proves nothing and never has, for there *can* be a unanimous belief in something that isn't so. The virtually universal opinion over thousands of years that the earth was flat, never flattened its spherical shape by one inch.

3. *The argument by belittlement.* Creationists frequently stress the fact that evolution is "only a theory." The impression this gives rise to is that a theory is an idle guess. A scientist, one gathers, arising one morning with nothing particular to do, decides that perhaps the moon is made of Roquefort cheese and instantly advances the Roquefort-cheese theory.

This is, of course, merely creationist naiveté. A theory (as the word is used by scientists) is a detailed description of some facet of the universe's workings that is based on long-continued observation and, where possible, experiment, that is the result of careful reasoning from those observations and experiments, and that has survived the critical study of scientists generally.

For example, we have the description of the cellular nature of living organisms (the "cell theory"), of objects attracting each other according to a fixed rule (the "theory of gravitation"), of energy behaving in discrete bits (the "quantum theory"), of light traveling through a vacuum at a fixed measurable velocity (the "theory of relativity"), and so on.

All are theories; all are firmly founded; all are accepted as valid descriptions of this or that aspect of the universe. They are not mere guesses, nor are they wild speculations. And no theory is better founded, more closely examined, more critically argued, and more thoroughly accepted than the theory of evolution. If it is "only" a theory, that is all it has to be.

Creationism, on the other hand, is *not* a theory. There is no evidence, in the scientific sense, that supports it — not one shred. Creationism, or at least the particular variety accepted by many Americans, is an expression of early Middle Eastern legend. It may be fairly described by those who wish to belittle it as "only a myth." Nor is that really belittlement for "only a myth" is exactly what creationism is.

4. *The argument from imperfection.* Creationists, in recent years, have stressed the "scientific" background of their beliefs. They point out that there are "scientists" who base their creationist beliefs on a careful study of geology, paleontology, and biology, and produce "textbooks" that embody those beliefs.

Virtually the whole "scientific" corpus of creationism, however, consists of the pointing out of imperfections in the evolutionary view. They insist that evolutionists can't show true transition states between species in the fossil evidence, that age-determinations through radioactive breakdown are uncertain, that alternate interpretations of this or that piece of evidence are possible, and so on.

Because the evolutionary view is not perfect and is not agreed upon in every detail by all scientists, creationists argue that evolution is false and that scientists, in supporting evolution, are basing their views on blind faith and dogmatism. (There, it must be admitted, creationists are on home territory. They have lived with blind faith and dogmatism from birth, and it is pleasant to see that they recognize it as an evil.)

The creationists are, to an extent, in the right here. The details of evolution are *not* perfectly known. Ever since Darwin first advanced his theory of the origin of species through natural selection, back in 1859, scientists have been adjusting and modifying Darwin's suggestions. After all, much has been learned about the fossil record, and about physiology, microbiology, biochemistry, ethology, and various other branches of life science in the past century and a quarter and it is to be expected that we can improve on Darwin. In fact, we *have* improved on him.

Nor is the process finished. It can never be, as long as human beings continue to question and to strive for better answers.

The details of evolutionary theory are in dispute precisely because scientists are *not* devotees of blind faith and dogmatism. They do not accept even as great a thinker as Darwin without question, nor do they hesitate to improve on him, nor do they accept *any* idea, new or old, without thorough argument. Even after accepting an idea, they stand ready to overthrow it if appropriate new evidence arrives.

If, however, we grant that a theory is imperfect and that details remain in dispute, does that disprove the theory as a whole?

Consider! I drive a car and you drive a car. I, for one, do not know *exactly* how an engine works. Perhaps you do not either. And it may be that our hazy and approximate ideas of the workings of an automobile are in conflict. Must we then conclude from this disagreement that an automobile does not run, or that it does not exist? Or, if our senses force us to conclude that an automobile does exist and run, that it is pulled by an invisible horse, since our engine-theory is imperfect?

However much scientists argue their differing beliefs in the details of evolutionary theory, or in the interpretation of the necessarily imperfect fossil record, they nevertheless firmly accept the evolutionary process itself.

Nor can imperfection in evolutionary theory possibly, in and of itself, lend credibility to creationism.

Suppose that one group of people held that the Empire State Building, by the evidence of their senses, was a skyscraper, while another group of people, pointing to an eighteenth-century description of the site, maintained that it was a Cape Cod cottage painted blue and white. If it turned out that the skyscraper devotees were uncertain as to whether the Empire State Building had an observation deck or not, that would not in and of itself prove that standing on the site was a Cape Cod cottage painted blue and white.

5. *The argument from distorted science.* Creationists have carefully learned enough of the terminology of science to attempt to disprove evolution by mouthing that terminology. They do this in numerous ways, but the most common example, at least in the mail I get, is the repeated assertion that the second law of thermodynamics demonstrates the evolutionary process to be impossible.

The second law of thermodynamics (expressed in kindergarten terms) states that all spontaneous change is in the direction of increasing disorder, that is, in a "downhill" direction. There can be no spontaneous build-up of the complex from the simple, therefore, for that would be moving "uphill." Clearly, then, so the creationist argument runs, since, by the evolutionary process, complex forms of life form from simple forms, that process, as described by scientists, defies the second law, and so creationism must be true.

This sort of argument implies that a fallacy clearly visible to anyone is somehow invisible to scientists, who must therefore be flying in the face of the second law through sheer perversity.

Scientists, however, *do* know about the second law and they are *not* blind. It's just that an argument based on kindergarten terms, as so many of the creationist arguments are, is suitable only for kindergartens.

To lift the argument a notch above the kindergarten level, the second law of thermodynamics applies to a "closed system," that is, to a system that does not gain energy from without or lose energy to the outside. The only truly closed system we know of is the universe as a whole.

*Within* a closed system, there are subsystems that can gain complexity spontaneously, provided there is a greater loss of complexity in another interlocking subsystem. The *overall* change is then a complexity-loss in line with the dictates of the second law.

Evolution *can* proceed and build up the complex from the simple, thus moving uphill, without violating the second law, as long as another interlocking part of the system — the sun, which delivers energy to the earth continuously — moves downhill (as it does) at a much faster rate than evolution moves uphill.

If the sun were to cease shining, evolution would stop and, indeed, so would life, eventually.

Unfortunately, the second law is a subtle concept that most people are not accustomed to dealing with, and it is not easy to see the fallacy in the creationist distortion. The fallacy becomes plainer, perhaps, if we consider the analogous treatment of another theory.

The theory of gravitation says, in kindergarten terms, that all objects in the earth's vicinity are attracted to the earth and, therefore, fall to the ground. Consequently, balloons and airplanes and rockets are clearly impossible.

— If you don't accept this, you needn't accept the creationists' kindergarten view of the second law of thermodynamics either.

There are many other "scientific" arguments used by creationists, some taking quite clever advantage of present areas of dispute in evolutionary theory, but every one of them is as disingenuous as the second-law argument.

The "scientific" arguments are organized into special creationist textbooks, which have all the surface appearance of the real thing and which school systems are heavily pressured to accept.

They are written by people who have not made any mark as scientists and, while they discuss geology, paleontology, and biology with correct scientific terminology, they are devoted almost entirely to raising doubts about the legitimacy of the evidence and the reasoning underlying evolutionary thinking, on the assumption that this leaves creationism as the only possible alternative.

Evidence actually in favor of creationism is not presented, of course, because none exists other than the word of the Bible, which it is current creationist strategy not to use.

6. *The argument from irrelevance.* Some creationists put all matters of scientific evidence to one side and consider all such things irrelevant. The Creator, they say, brought life, and the earth, and the entire universe into being ten thousand years or so ago, complete with all evidence for an eons-long evolutionary development. The fossil record, the decaying radioactivity, the receding galaxies, were all created as they are and the evidence they present is an illusion.

Of course, this argument is itself irrelevant, for it can be neither proved nor disproved. It is not an argument, actually, but a statement. I can say that the entire universe was created two minutes ago, complete with all its history books describing a nonexistent past in detail, and with all the persons now alive equipped with full memories; you, for instance, in the process of reading this article in midstream with a memory of what you had read in the beginning — which you had not really read.

That, too, can be neither proved nor disproved.

Ask yourself, though, what kind of a Creator would produce a universe containing so intricate an illusion.

It would mean that the Creator formed a universe that contained human beings whom he had endowed with the faculty of curiosity and the ability to reason. He supplied those human beings with an enormous amount of subtle and cleverly self-consistent evidence designed to mislead that curiosity and that reasoning ability and cause them to be convinced that the universe was created fifteen billion years ago and developed by evolutionary processes that included the creation and development of life on earth.

Why?

Does the Creator take pleasure in misleading us? Does it amuse him to watch us go wrong? Is it part of a test to see if human beings will deny their senses and their reason in order to cling to a myth? Is it to give him an excuse to consign us all to hell for not denying our senses and our reason?

Can it be that the Creator is a cruel and malicious prankster, with a vicious and adolescent sense of humor?

If so, it might be just as well if the creationists were honest, and said so.

7. *The argument from authority.* The Bible says that God created the world in six days, and the Bible is the inspired word of God.

To the average creationist this is all that counts, really. All other arguments are merely a tedious way of countering the propaganda of all those wicked humanists, agnostics, and atheists who are not satisfied with the clear word of the Lord.

To be sure, the creationist leaders are careful not to use that argument because that would make their point of view a religious one and they would not be able to get it into our secular school-system. They have to borrow the clothing of science, no matter how badly it fits them and no matter how grotesque it makes them appear, in order to call themselves "scientific" creationists. They must also be careful to speak only of a "Creator" and never mention that this Creator happens to be the God of the Bible. The careful impression is left that he might, for all anyone knows, be Moloch or Chemosh or any of the other heathen abominations the Bible speaks of.

We cannot, however, take this sheep's clothing seriously. However much the creationist leaders might hammer away at their "scientific" and "philosophical" points, they would be helpless and a laughing stock if that were all they had.

It is religion, the simple fervor of medieval piety, that recruits their squadrons. Tens of millions of Americans, who neither know or understand the actual arguments for, or even against, evolution, march in the Army of the Night with their Bibles held high. And they are a strong and frightening force, impervious to and immunized against the feeble lance of mere rationality.

But let us move on. Even if I am right and the evolutionists' case is very strong, have not creationists, whatever the emptiness of their case, a right to be heard?

If their case is empty, isn't it perfectly safe to discuss it, since the emptiness would then be apparent? Wouldn't it be *best* to discuss it, so that the emptiness could be displayed?

Why, then, are evolutionists so reluctant to have creationism taught in the public schools on an equal basis with evolutionary theory? Can it be that the evolutionists are not as confident of their case as they pretend? Are they afraid to allow youngsters a clear choice?

In this connection, there are two points to be made.

First, the creationists are somewhat less than honest in their demand for equal time. It is not *they* who are repressed, for schools are by no means the only place in which the dispute between creationism and evolutionary theory is played out.

There are the churches, for instance, which are a much more serious influence on most Americans than the schools are. To be sure, many churches are quite liberal, have made their peace with science, and find it easy to live with scientific advance — even with evolution. But the majority of the less modish and citified churches are bastions of creationism.

The influence of the church is naturally felt in the home, in the newspapers, and in all of surrounding society. It makes itself felt in the nation as a whole, even in religiously liberal areas, in ten thousand subtle ways, in the nature of holiday observance, in expressions of patriotic fervor, even in total irrelevancies. Thus, in 1968, a team of astronauts circling the moon were instructed to read the first few verses of Genesis, as though NASA felt it had to placate the public lest they rage against the violation of the firmament. At the present time, even the current president of the United States has expressed his creationist sympathies.

It is *only* in school that American youngsters in general are ever likely to hear any reasoned exposition of the evolutionary viewpoint. They may find such a viewpoint in books or even, on occasion, on television; but church and family can easily censor books and television, and only the school is beyond their control.

But only just barely beyond. Even though schools are now allowed to teach evolution, teachers are bound to be apologetic about it, knowing full well their jobs are at the mercy of school boards not noted for intellect or for their breadth of scientific view.

Then, too, in schools, students are not required to believe what they learn about evolution — merely to parrot it back on tests. If they fail to do so, their punishment is nothing more than the loss of a few points on a test or two.

In the creationist churches, however, the congregation is required to *believe* under the threat of hellfire. Impressionable youngsters, taught to believe that they will go to Hell if they listen to the evolutionary doctrine, are not likely to listen in comfort, or to believe if they do.

Well, then, creationists, who control the church and the society they live in, and who face the public school as the only place where evolution is even briefly mentioned in a possibly favorable way, find they cannot stand even so minuscule a competition and demand "equal time."

Do you suppose their devotion to "fairness" is such that they will give equal time to evolution in their churches? You know they won't. What's theirs is theirs. What's yours is negotiable.

Second, the real danger is the manner in which creationists want their "equal time."

In the scientific world, there is free and open competition of ideas, and even a scientist whose suggestions are not accepted is nevertheless free to continue to argue his case.

In this free and open competition of ideas, creationism has clearly lost. It has been losing, in fact, since the time of Copernicus three and a half centuries ago.

Creationism refuses to accept the decision, placing myth above reason, and is now calling in the power of the government. They want the government to *force* creationism into the schools against the verdict of the free and open competition of ideas. Teachers must be *forced* to present creationism as though it has equal intellectual respectability with evolutionary doctrine.

What a precedent this sets!

If the government can mobilize its policemen and its prisons to make certain that teachers give creationism equal time, they can next use force to make sure that teachers declare creationism the victor so that evolution may be evicted from the classroom altogether.

We will have established the full groundwork, in other words, for barbarism, for legally enforced ignorance, and for totalitarian thought-control.

And what if the creationists win? They might, you know, for there are millions who, faced with the choice between the Bible and science, will choose the Bible and reject science, regardless of the evidence.

This is not entirely because of a traditional and unthinking reverence for the literal words of the Bible; there is also a pervasive uneasiness, or actual fear, of science, that will drive even those who care little for religion into the arms of the creationists.

For one thing, science is uncertain. Theories are subject to revision; observations are open to a variety of interpretations, and scientists quarrel among themselves. This is disillusioning for those untrained in the scientific method, and these people tend to turn to the rigid certainty of the Bible as presented by its thumpers. There is something comfortable about a view that allows for no deviation and that spares you the painful necessary of having to think.

Second, science is complex and chilling. The mathematical language of science is understood by very few. The vistas it presents are scary — an enormous universe ruled by chance and impersonal rules, empty and uncaring, ungraspable and vertiginous. How comfortable to turn instead to a small world, only a few thousand years old, and under God's personal and immediate care; a world in which you are His peculiar concern and where He will *not* consign you to Hell if you are careful to follow every word of the Bible as interpreted for you by your television preacher.

Third, science is dangerous. There is no question but that such products as poison gas, nuclear weapons and power stations, and genetic engineering are terrifying. It may be that civilization is falling and the world we know is coming to an end. In that case, why not turn to religion and look forward to the Day of Judgment, in which you and your fellow-believers will be lifted into eternal bliss, and have the added joy of watching the scoffers and disbelievers writhe forever in torment.

So why might they not win?

Spain dominated Europe and the world in the sixteenth century, but in Spain orthodoxy came first and all divergence of opinion was ruthlessly suppressed. The result was that Spain settled back into blankness and did not share in the scientific, technological, and commercial ferment that bubbled up in other nations of Western Europe. Spain remained an intellectual backwater for centuries.

In the late seventeenth century, France in the name of orthodoxy revoked the Edict of Nantes and drove out many thousands of Huguenots, who added their intellectual vigor to lands of refuge like Great Britain, the Netherlands, and Prussia, while France was permanently weakened.

In more recent times, Germany hounded out the Jewish scientists of Europe. These, arriving in the United States, added immeasurably to scientific advance here, while Germany lost so heavily that there is no telling how long it will take it to regain its former scientific eminence. The Soviet Union, in its fascination with Lysenko, destroyed its geneticists, and set back its biological sciences for decades. China, during the Cultural Revolution, turned against Western science and is still laboring to overcome the devastation that resulted.

Are we now, with all these examples before us, to ride to destruction under the same tattered banner of orthodoxy? With creationism in the saddle, American science will wither, and we will raise a generation of ignoramuses who will not be equipped to run the industry of tomorrow, much less to generate the new advances of the days after tomorrow.

We will inevitably recede into the backwater of civilization, and those nations that retain open scientific thought will take over the leadership and the cutting edge of human advance.

I don't suppose that the creationists really plan the decline of the United States, but their loudly expressed patriotism is as simple-minded as their "science" and, if they win out, they will, in their folly, achieve the opposite of what they say they wish.

*Afterword*: You will find my thesis restated in the next essay, more briefly and in sprightlier language, for I was writing for *Penthouse* this time, rather than the *New York Times*. If you're curious about how I handled it the second time around, read on. Otherwise, you might skip the next essay and come back to it later. Go ahead, you won't hurt my feelings.

# 2

# Creationism and the Schools

There has been an outcry in this nation in recent years, by spokesmen of the radical right-wing, to the effect that something they call "creationism" be taught in tax-supported public schools. They want this presented on an equal basis with the concept of evolution as an explanation of the origin of the universe, of life, of human beings.

It *seems* fair. Why shouldn't both sides have the same chance? Why should the evolutionists oppose the teaching of creationism?

Yet equal time for both views is *not* fair. It is pernicious.

The concepts of evolution and creationism are *not* equal. The evolutionary view has been built up painstakingly over a period of two centuries on the basis of scientific study, and it has behind it an enormous body of evidence and reasoning. All biologists, of any reputation at all, accept the evidence that present-day species have developed slowly from simpler forms; that the unit of life, the cell, has developed from pre-cellular scraps of life; and that these, in turn, have arisen from nonliving materials by changes that are in accord with the laws of nature over a vast stretch of several billions of years.

The exact mechanism of evolution, the fine details, remain under dispute, since the process of discovery and development is not yet done and may never be entirely done. Even the most argumentative of those who quarrel over the details do not, however, deny the evolutionary concept itself.

Creationists, on the other hand, present no evidence in favor of their view. They argue entirely from the negative. They maintain that if the concept of evolution is found wanting, then that alone is sufficient to force acceptance of creationism.

They then insist that the concept of evolution is indeed found wanting.

16

They point out insufficiencies, contradictions, and uncertainties in the evolutionary arguments and say, triumphantly, "Thus we establish creationism!"

And yet, in the first place, the insufficiencies they present are often advanced in distorted, simplistic, and downright erroneous ways. In the second place, some of those insufficiencies are matters over which biologists are indeed undecided, but which affect merely the details of mechanism and not the concept of evolution itself.

And in the third place, even if the concept of evolution were indeed insufficient, that would not, of itself, prove the validity of the concept of the independent production of each species by a "Creator." Other alternatives may exist and the choice among them would have to rest on positive evidence. Thus, if a close investigation were to show that our notions of reproductive physiology were not entirely right, that would not, of itself, prove that babies were brought by the stork. They might, indeed, have been found under cabbage leaves, or have been delivered in the doctor's little black bag.

In order to establish creationism as a rational concept, the creationists must advance scientifically valid evidence *for* their beliefs, and not merely try to poke holes in other views. They cannot simply question whether the universe is really fifteen billion years old by casting doubt on the Hubble constant. They must present reasonable evidence that the universe *is,* in fact, ten thousand years old (or whatever figure they would like to maintain). Needless to say, this they have never done.

For these reasons, creationism has never established itself in the one place that really counts — in the marketplace of scientific ideas.

Science is a self-correcting process, and scientists *do* change their views; but they do so only on the basis of new evidence or of a new and convincing presentation of a line of reasoning. Scientists refused to accept the notion of drifting continents on the basis of evidence advanced in 1913 and thereafter. New evidence was obtained in the 1960s, and a revised and improved version of the concept was then accepted with surprising speed.

It is possible that the day may come when evolution will indeed turn out to be insufficient and when new evidence in favor of creationism will force a change of view, but that day *has not yet come.* Nothing the creationists say bears promise that it will ever come and, since that is so, it is impossible, scientifically, to ask that creationism be taught in the schools today as a reasonable alternative to evolution.

The fact that some people earnestly believe in creationism is insufficient. The existence of that belief is a legitimate matter of interest in courses of history, sociology, and psychology, and in those courses creationism may well be discussed in detail, but it doesn't belong in science.

But suppose we *were* to teach creationism. What would be the content of the teaching? Merely that a Creator formed the universe and all species of life ready-made? Is that it? Nothing more? No details?

American creationists seem to accept the biblical tale of creation, but is that the only pattern of creation possible? Millions of people the world over who believe in a divine creator of some sort do not accept the Bible as a holy book.

In fact, many people who read the Bible disagree on the manner of interpretation of its account of the creation. They may accept the account as poetry, as allegory, as symbolism; they may see in it deep ethical and moral meanings — but they don't accept it as a *literal* description of how the universe began.

What, then, do we teach if we teach creationism? Which view do we accept? Do we try to choose among them on the basis of scientific evidence? Do we just teach them all on an equal basis? If creationists simply want the literal words of the Bible taught, then that is manifestly unfair to all the competing creationist notions.

It might be possible to argue that, if creationism is so empty of content and so transparently unscientific, there is certainly no harm in offering it as an alternative. Surely no one would accept it. Some people even argue that, if scientists object to "equal time," they must not really have a good case.

Ah, but it is not equal time the creationists want. That little slogan is merely the smile of the crocodile.

School is not the only place where the origins of life and the universe are dealt with. There are also those churches that have creationist views (not all churches, by any means, of course). In those churches, *only* the creationist view is presented. There is no question of "equal time" there. Children are therefore exposed only to creationist views there, and in their homes, for many years before they hear of evolution in the schools. And they are threatened with hellfire if they doubt. Where is the "equal time"?

The teaching of evolution in the public schools is a very recent thing. It was not many decades ago when in the strongholds of creationism the teaching of evolution was forbidden. That was what the Scopes "monkey trial" was all about. Scopes had mentioned evolution in class and that was a *crime*. Where was the "equal time" then?

Even now, the teaching of evolution in public schools is not a very strong affair. In many areas of the nation, people of creationist views heckle school boards, school principals, and school teachers to the point where, if evolution is mentioned at all, it is done in an apologetic whisper. The creationists attempt to ride herd on the libraries, too, and do their best to pull out every book that doesn't suit them.

And they want "equal time"? Don't kid yourself. They want all the time there is. One can see why, too. Their case is so weak, so nonexistent, in fact, that the only way they can feel sure of maintaining it is to have their victims *never hear of anything else.*

Yet none of what I have so far said reaches the real deadliness of the situation.

Creationist views, after all, continue to be firmly rejected in the marketplace of scientific ideas. There can be no other way as long as creationist views are so empty of content.

So the creationists call on the government. They browbeat legislators and executives and insist on *laws* defining what is scientifically valid and dictating what is to be taught.

What a dangerous precedent this is! If the Supreme Court can be bullied into declaring such things constitutional, it would go a long way toward putting an end to pluralism in this country, and to free thought. We would be on the road to an established church and an official orthodoxy.

All historical precedents show that the ability to censor and to enforce orthodoxy is a delight that knows no limits. Today "equal time," tomorrow the world. Today it is your views on science, tomorrow the way you dress and speak and behave.

It is not merely creationism that we are fighting in this matter. Behind it are the old enemies of bigotry and darkness, and we must not complain about this endless battle. The price of liberty, said Jefferson, is eternal vigilance.

---

# 3

# The Reagan Doctrine

Some time ago, Ronald Reagan pointed out that one couldn't trust the Soviet government because the Soviets didn't believe in God and in an afterlife and therefore had no reason to behave honorably but would be willing to lie and cheat and do all sorts of wicked things to aid their cause.

Naturally, I firmly believe that the president of the United States knows what he is talking about, so I've done my very best to puzzle out the meaning of that statement.

Let me begin by presenting this "Reagan Doctrine" (using the term with all possible respect): "No one who disbelieves in God and in an afterlife can possibly be trusted."

If this is true (and it must be if the president says so), then people are just naturally dishonest and crooked and downright rotten. In order to keep them from lying and cheating everytime they open their mouths, they must be bribed or scared out of doing so. They have to be told and made to believe that, if they tell the truth and do the right thing and behave themselves, they will go to Heaven and get to plunk a harp and wear the latest design in halos. They must also be told and made to believe that if they lie and steal and run around with the opposite s-x, they are going to Hell and will roast over a brimstone fire forever.

It's a little depressing, if you come to think of it. By the Reagan Doctrine, there is no such thing as a person who keeps his word just because he has a sense of honor. No one tells the truth just because he thinks that it is the decent thing to do. No one is kind because he feels sympathy for others, or treats others decently because he likes the kind of world in which decency exists.

Instead, according to the Reagan Doctrine, anytime we meet someone who pays his debts, or hands in a wallet he found in the street, or stops to help a blind man cross the road, or tells a casual truth—he's just buying himself a ticket to heaven, or else canceling out a demerit that might send him to Hell. It's all a matter of good, solid business practice; a matter of turning a spiritual profit and of responding prudently to spiritual blackmail.

Personally, I don't think that any of us—I, or you, or even President Reagan—would knock down an old lady and snatch her purse the next time we're short a few bucks, if only we were sure of that heavenly choir, or if only we were certain we wouldn't get into that people-fry down in Hell. But by the Reagan Doctrine, if we didn't believe in God and in an afterlife, there would be nothing to stop us, so I guess we all would.

But let's take the reverse of the Reagan Doctrine. If no one who disbelieves in God and in an afterlife can possibly be trusted, it seems to follow that those who *do* believe in God and in an afterlife *can* be trusted.

Since the American government consists of God-fearing people who believe in an afterlife, it seems pretty significant that the Soviet Union nevertheless would not trust us any farther than they can throw an ICBM. Since the Soviets are slaves to Godless communism, they would naturally think everyone else is as evil as they are. Consequently, the Soviet Union's distrust of us is in accordance with the Reagan Doctrine.

Yet there are puzzles. Consider Iran. The Iranians are a God-fearing people and believe in an afterlife, and this is certainly true of the mullahs and ayatollahs who comprise their government. And yet we are reluctant to trust them for some reason. President Reagan himself has referred to the Iranian leaders as "barbarians."

Oddly enough, the Iranians are also reluctant to trust us. They referred to the ex-president (I forget his name, for he is never mentioned in the media anymore) as the "Great Satan" and yet we all know that the ex-president was a born-again Christian.

There's something wrong here. God-fearing Americans and God-fearing Iranians don't trust each other and call each other terrible names. How does that square with the Reagan Doctrine?

To be sure, the God in whom the Iranians believe is not quite the God in whom we believe, and the afterlife they believe in is a little different from ours. There are no houris, alas, in our Heaven. We call our system of belief Christianity and they call theirs Islam, and come to think of it, for something like twelve centuries, good Christians believed Islam was an invention of the Devil and believers in Islam ("Moslems") courteously returned the compliment so that there was almost continuous war between them. Both sides considered it a holy war and felt that the surest way of going to Heaven was to clobber an infidel. What's more, you didn't have to do it in a fair and honorable way, either. Tickets of admission just said, Clobber!

This bothers me a little. The Reagan Doctrine doesn't mention the variety of God or afterlife that is concerned. It doesn't indicate that it matters what you call God—Allah, Vishnu, Buddha, Zeus, Ishtar. I don't think that President Reagan meant to imply a Moslem couldn't trust a Shintoist or that a Buddhist couldn't trust a Parsee. I think it was just the Godless Soviets he was after.

Yet perhaps he was just being cautious in not mentioning the fact that the variety of deity counted. But, even if that were so, there are problems.

For instance, the Iranians are Moslems and the Iraqi are Moslems. Both are certain that there is no God but Allah and that Mohammed is his prophet and believe it with all their hearts. And yet, at the moment, Iraq doesn't trust Iran worth a damn, and Iran trusts Iraq even less than that. In fact, Iran is convinced that Iraq is in the pay of the Great Satan (that's God-fearing America, in case you've forgotten) and Iraq counters with the accusation that it is Iran who is in the pay of the Great Satan. Neither side is accusing the Godless Soviets of anything, which is a puzzle.

But then, you know, they *are* Moslems and perhaps we can't just go along with any old god. I can see why Reagan might not like to specify, since it might not be good presidential business to offend the billions of people who are sincerely religious but lack the good taste to be Christians. Still, just among ourselves, and in a whisper, perhaps the only people you can really trust are good Christians.

Yet even that raises difficulties.

For instance, I doubt that anyone can seriously maintain that the Irish people are anything but God-fearing, and certainly they don't have the slightest doubts concerning the existence of an afterlife. Some are Catholics and some are Protestants, but both of these Christian varieties believe in the Bible and in God and in Jesus and in Heaven and in Hell. Therefore, by the Reagan Doctrine, the people of Ireland should trust each other.

Oddly enough, they don't. In Northern Ireland there has been a two-sided terrorism that has existed for years and shows no sign of ever abating. Catholics and Protestants blow each other up every chance they get, and there seems to be no indication of either side trusting the other even a little bit.

But then, come to think of it, Catholics and Protestants have had a thing about each other for centuries. They have fought each other, massacred each other, and burned each other at the stake. And at no time was this conflict fought in a gentlemanly, let's-fight-fair manner. Any time you caught a heretic or an idolator (or whatever nasty name you wanted to use) looking the other way, you sneaked up behind him and bopped him and collected your ticket to Heaven.

We can't even make the Reagan Doctrine show complete sense here in the United States. Consider the Ku Klux Klan. They don't like the Jews or the Catholics, but then, the Jews don't accept Jesus and the Catholics do accept the pope, and these fine religious distinctions undoubtedly justify distrust by a narrow interpretation of the Reagan Doctrine. The Protestant Ku Klux Klan can only cotton to Protestants.

Blacks, however, are predominantly Protestant, and of southern varieties, too, for that is where their immediate ancestors learned their religion. Ku Kluxers and blacks have very similar religions and therefore even by a narrow interpretation of the Reagan Doctrine should trust each other. It is difficult to see why they don't.

What about the Moral Majority? They're absolute professionals when it comes to putting a lot of stock in God and in an afterlife. They practice it all day, apparently. Naturally, they're a little picky. One of them said that God didn't listen to the prayers of a Jew. Another refused to share a platform with Phyllis Schlafly, the Moral Majority's very own sweetheart, because she was a Catholic. Some of them don't even require religious disagreements, just political ones. They have said that one can't be a liberal and a good Christian at one and the same time; so if you don't vote right, you are going straight to Hell whatever your religious beliefs are. Fortunately, at every election they will tell you what the right vote is so that you don't go to Hell by accident.

Perhaps we shouldn't get into the small details, though. The main thing is that the Soviet Union is Godless and, therefore, sneaky, tricky, crooked, untrustworthy, and willing to stop at nothing to advance their cause. The United States is a God-fearing and therefore forthright, candid, honest, trustworthy, and willing to let their cause lose sooner than behave in anything but the most decent possible way.

It bothers the heck out of me therefore that there's probably not a country in the world that doesn't think the United States, through the agency of the CIA and its supposedly underhanded methods, has upset governments in Guatemala, Chile, and Iran (among others), has tried to overthrow the Cuban government by a variety of economic, political, and even military methods, and so on. In every country, you'll find large numbers who claim that the United States fought a cruel and unjust war in Vietnam and that it is the most violent and crime-ridden nation in the world.

They don't seem to be impressed by the fact that we're God-fearing.

Next they'll be saying that Ronald Reagan (our very own president) doesn't know what he's talking about.

# 4

# The Blind Who Would Lead

In the United States, an old, old phenomenon resurfaced in 1980. It is the voice of self-righteous, all-knowing, narrow-minded "religion," this time in the form of the self-styled "Moral Majority," which has as its objectives the punishment of politicians who deviate from M.M. principles, and the dictation to all Americans of what they should read, think, and believe.

The Moral Majority speaks with the voice of absolute authority.

This is not to be confused with the kind of authority expressed by scientists, who can only claim to hope they are right pending further information. The greatest scientists have been wrong on this point or that—Newton on the nature of light, Einstein in his views of the uncertainty principle—and this does not lessen respect for their achievements. Scientists expect to be improved on and corrected; they hope to be. Science has its "authority," but it is an open and nonauthoritarian authority.

The Moral Majority, however, speaks, it would seem, with the voice of God. How do we know they do? Why, they themselves say so; and, since they speak with the voice of God, then their testimony that they do so is unshakable. Q. E. D. And since the voice of God is never wrong and cannot be wrong (the Moral Majority, speaking with that voice, says so) any spokesman of the M.M. is never wrong and cannot be wrong.

The Moral Majority is, in other words, a closed intellectual system, without possibility of change or admission of error. It insists that all the answers exist and have existed from the beginning because God wrote them all out in the Bible and we need only observe them to the letter.

Surely this puts an end forever to any hope of social or intellectual advance or to any rational adaptation to changing conditions. For what does the Bible say? "The letter killeth" (Cor. II, 3:6).

24

We have had thousands of years of experience with the kind of absolute self-anointed authority that little men adopt and call "religion." We have watched the Christian nations of the world fight each other for many centuries, each believing itself under the peculiar protection of a God they all insist is universal. Each prays separately for the destruction of the enemy and praises jubilantly God's assistance in helping them bring death and misery to that enemy, although both sides, presumably, are equally God's children.

(Nor is it only the Christians. We see Moslem Iraq fighting Moslem Iran, with each insisting that the universal Allah is aiding only its own side while the United States is supporting the other.)

One can only draw the conclusion that the "religionists" (to be distinguished from the wisely, ethically, and universally religious) are convinced, in each nation, that God is himself a member of that nation and is rather proud of being so.

Certainly I suspect that the Americans of the Moral Majority take it for granted that God is an American citizen (naturalized, of course) and that he is, moreover, a member of the conservative wing of the Republican Party and voted for Ronald Reagan.

Furthermore, the Moral Majority types are convinced that God is forever intervening in human affairs in a kind of wrathful way but that he is fortunately weak-willed. He may send a drought, for instance, to punish sinners; but if everyone prays and pleads and wails, he'll say, "Oh, all right," and let it rain.

If there is an earthquake and a thousand people die, and one person is uncovered from a ruined house, unhurt, the Moral Majority types cry "A miracle!" and fall to their knees in gratitude. And the thousand who died—whose deaths, indeed, were necessary to convert the one survivor into a miracle, what of them?

Not to worry. If anything really puzzling happens, it is only necessary to remind everyone that merely human minds cannot expect to penetrate the deep, mysterious purposes of God. Except that the Moral Majority does it all the time, when they want to whip and harry the rest of us.

The Moral Majority feels absolutely secure in being under the protection of an American Republican conservative God. Still—just the same—they favor a strong national defense. God will certainly destroy the Godless Soviets, but he'll need a lot of very advanced bombers, missiles, and nuclear bombs to do it with. (Scientists are perfectly moral people, to the Moral Majority, as long as they design sophisticated war weapons to control the population problem by raising the death rate—as opposed to any evil attempt they make to control it by lowering the birth rate.)

What's more, the Moral Majority types do not need to study science or consider its observations and conclusions to know that those observations are misleading and those conclusions are not only wrong—but deeply wicked.

The M.M. has its own textbook of biology, astronomy, and cosmogony, in the form of the Bible, a collection of two-thousand-year-old writings by provincial tribesmen with little or no knowledge of biology, astronomy, and cosmogony.

To be sure, the Bible contains the direct words of God. How do we know? The Moral Majority says so. How do *they* know? They *say* they know and to doubt it makes you an agent of the Devil or, worse, a L-b-r-l D-m-cr-t.

And what does the Bible textbook say? Well, among other things it says the earth was created in 4004 B.C. (Not actually, but a Moral Majority type figured that out three and a half centuries ago, and his word is also accepted as inspired.) The sun was created three days later. The first male was molded out of dirt, and the first female was molded, some time later, out of his rib.

As far as the end of the universe is concerned, the Book of Revelation (6:13–14) says: "And the stars of heaven fell unto the earth, even as a fig tree casteth her untimely figs, when she is shaken of a mighty wind. And the heaven departed as a scroll when it is rolled together."

The Bible textbook, then, says that the sky is a thin sheet of something or other that can be rolled up in the same way a scroll of parchment can be rolled up, and that the stars are little dots of light that can be shaken off that scroll and allowed to fall to the earth.

Imagine the people who believe such things and who are not ashamed to ignore, totally, all the patient findings of thinking minds through all the centuries since the Bible was written.

And it is these ignorant people, the most uneducated, the most unimaginative, the most unthinking among us, who would make of themselves the guides and leaders of us all; who would force their feeble and childish beliefs on us; who would invade our schools and libraries and homes in order to tell us what books to read and what not, what thoughts to think and what not, what conclusions to accept and what not.

And what does the Bible say? "If the blind lead the blind, both shall fall into the ditch" (Mat. 15:14).

# 5

# Creeping Censorship

It takes a calm sense of security to be willing to let each person have his say in speaking and writing, regardless of the content thereof. The security may arise out of the conviction that in the free conflict of ideas, truth and sanity will prevail, even if that takes a little time, or out of an abstract devotion to free thought, whatever the outcome.

The trouble is that few people have the strength to cling to this conviction if it appears to them that some view, dangerous to their beliefs or to themselves, is gaining ground. For that reason, the impulse to censorship is insidious in its origin, arising among good and honest people, for reasons that seem irreproachable.

People with some sense of decorum find blatant pornography unpleasant and even disgusting and fear its effect on the young, the uneducated, and the mentally unstable. People with firm belief in traditional values are distressed at iconoclastic views expressed with eloquence and conviction. People who consider themselves oppressed fear views that they interpret as designed to continue or intensify that oppression.

In every case, it seems reasonable to suppress or circumscribe the free expression of ideas; and it is exceedingly difficult for many, in the name of abstract freedom, to condone blatant displays of sexual perversity or to defend the right to express views that can only be described as cruel and ignorant bigotry.

And, yet, however laudable and respectable the first steps toward censorship may appear, they must be viewed with fear and trepidation. Censorship feeds upon itself. Once it is established as a legitimate governmental activity, the pressure is always toward a broadening of the shutdown. Wipe out the more blatant examples of pornography, and of what remains some will seem worse than others and there will be an outcry against those. And if all are shut down, there will be arguments as to what might be pornographic by implication, even though nothing is explicit, and the tendency will increasingly be to play it safe by forbidding anything on the borderline — and then on the new borderline, and so on.

All this would also be true of censorship in the name of traditional values or national security. The more one gets into the habit of censorship, the more one views with suspicion what is left. The more one blots out dissent, the clearer the view that the nation and its morals are in danger (else why the blotting out) and the more respectable and patriotic it is to extend the erasure.

Remember, too, that there are always those who must make the decision as to what to censor and what not. If they censor too much, that will scarcely endanger them, for that which is censored is hidden from view and few are aware of its existence. If they censor too little, however, that uncensored material remains on view, and the ultrarespectable and ultrapatriotic might raise a hue and cry that would endanger the censor's job — or life. It is certain, then, that censors would do more rather than less, and that censorship would grow and spread like an infectious disease.

We see the beginnings of censorship in our society now. Led by the radical right, self-appointed vigilantes descend upon schools and libraries to force books off the shelves — because they possess naughty words, or naughty ideas, or simply because they run counter to some belief held by these vigilantes.

If these amateur censors have their way, then because *they* don't like a book (such as *Huckleberry Finn,* which, by common consent, is the greatest American novel) then first no child would be allowed to read it in school, and next, I'm sure, no child would be allowed to read it anywhere, and finally, I'm sure, no person of whatever age would be allowed to read it anywhere at anytime.

And from books, they will pass on to songs, to speech, to thought, and will create an America (if they can) in their own image.

Imagine an America cast in that image, the image of the Moral Majority, superpatriotic, superrespectable — with only certain narrow points of view defined as either patriotic or respectable. Imagine a society gray and humorless, with views that are identical from sea to shining sea, and with nothing otherwise permitted. Imagine an America compressed — narrow, single-viewed, ignorant. There have been many such societies in history.

It won't happen here? Of course not, if the movement toward it is stopped; but the farther it advances the harder it is to stop, and the best time to stop it is at the beginning — *now!*

# —6—

# Losing the Debate

Every once in a while, some scientist who accepts the view that the universe, life, and human beings have developed slowly over billions of years through evolutionary processes is lured into a debate with a "creationist" who insists that the universe, life, and human beings have been brought into existence only a few thousand years ago, in just about its present form, by supernatural action.

To serious students of science, it would seem that a scientist must win such a debate. After all, on the side of the scientist are vast numbers of all kinds of observations, to say nothing of careful argument and unassailable logic. On the side of the creationist, there is, from the scientific point of view, exactly nothing.

And yet, somehow, in such debates, the creationist often appears to have it all his own way, while the scientist is reduced to an ineffective defense. Why is that?

No mystery! The scientist has generally spent his professional life in scientific debate with other scientists. The weapons in such debates are evidence and careful reasoning. Opposing points of view are maintained unemotionally, and all participants follow the rules of the scientific method. If one or all of those taking part in a scientific debate are not good speakers, that does not matter very much. It is the content that counts.

The creationist, however, is often a showman, and usually a polished speaker. He has no concern for scientific evidence or careful reasoning and is on the stage in order to win debating points with the audience. He sounds much better than the scientist as a matter of course. What he says is worth nothing, but it invariably *sounds* good. The scientist usually is untrained in handling such showman-tactics and cannot respond effectively.

The scientist, moreover, is conditioned to admit uncertainty and ignorance. That is an essential part of science. The creationist, therefore, attacks in that direction. He points out places in the evolutionary view where there are uncertainties and confusion, and the scientist must, perforce, admit it. He is forced to defend and explain endlessly.

The scientist, in fact, once maneuvered into the defense, almost never thinks of shifting to the attack. He never demands the actual *evidence* for the view that there was a universal creation a few thousand years ago. The creationist is never forced to state whether many men and women were then created, or only one pair; whether both sexes were created at once, or women after men; and whether serpents could at one time speak.

What's more, there is almost always a built-in bias on the part of the audience. Almost invariably, the debate takes place before people who are only sketchily trained in science, if at all, and who have, in many cases, an automatic reverence for the literal words of the Bible.

The creationist seems to be on the side of the Bible and religion (and Mom, and baseball, and apple pie, too), while the scientist is easily represented as being against these things. The audience, therefore, tends to place itself clearly behind the creationist, and that further confuses and demoralizes the scientist.

What ought a scientist to do then?

It seems to me he ought to decline to debate these showmen on their terms if he lacks the talent for the rough-and-tumble. And if he thinks he has the talent, he should not bother defending evolution; he should move to force his opponent to present the evidence for creationism. Since there isn't any, the results could be humorous.

# Part II

# Other Aberrations

# 7

# The Harvest of Intelligence

A sperm bank has been set up. The sperm of men of proven brilliance (Nobel Prize winners, for instance) are frozen and set to one side, awaiting the appearance of women of proven brilliance who wish to be artificially inseminated by these supersperm in order to produce unusual infants of double brilliance.

If enough men contribute sperm and enough women are willing and qualified to accept them, the vision is of a whole group of double-brilliants— a group, one might think, that the human species sorely needs.

And yet the only reason I can think of to hope that the experiment continues and is brought to fruition is my conviction that it will prove to the world that this sort of thing won't work.

The human brain is the most complex bit of organized matter we know of anywhere in the universe. It is incredibly more complex than a star, for instance, and that's why we know so much more about stars than about the human brain.

And the most complex aspect of the human brain is its intelligence; so it's not surprising that we know next to nothing about that. We aren't even certain we know what we are measuring when we say we are measuring intelligence, or just exactly what it is that the "intelligence quotient" is supposed to be.

We know how inheritance works in its simpler aspects. We know about genes, and we can work out very nicely what happens to characteristics that are controlled by single genes in a very straightforward manner. We know the inheritance characteristics of plant-stem lengths, and plant-seed skins, and insect eye-color, human eye-color, and human blood agglutinogens, to name a few.

We don't know the relationships of genes and intelligence, however. We don't know how many genes play a role in the development of intelligence, how the different genes interrelate, the nature of the chemistry they control, or anything else. There is therefore no way in which we can predict that the offspring of two intelligent parents will be intelligent, in the same way that we can predict that the offspring of two blue-eyed parents will be blue-eyed.

All that can be done is to make statistical studies of IQs of parents and children. The trouble is that we're not sure how good the IQ tests are, or whether they're any good at all. We're not reasonably *sure*. And we don't know how much is the effect of chance, or of nongenetic influences.

For instance, all sperm cells of a given male should, in theory, have identical sets of genes, but it is quite possible that no two are exactly alike. Genes don't always duplicate themselves exactly as millions of different sperm cells are formed. There might be a large range in quality of the sperm cells produced by a particular male at a particular time, and it is just a matter of chance which sperm fertilizes a particular egg (and every egg cell formed by a woman can be different, too, and likely is).

If it's a matter of eye-color, that is simple and tiny gene-variations may not matter; but, in something as subtle and delicate as human intelligence, chance may be everything.

Then, too, what of the influence of the mother's physiology and chemistry during those nine months when a fetus is growing; when it is fed and cared for by way of the mother's bloodstream and placenta? Imperfect housing characteristics (so to speak) may stunt or distort the fetus's growth, and even a delicate and subtle distortion, unnoticeable in other ways, might affect so delicate and subtle a thing as intelligence. Remember, too, that a very intelligent and admirable woman may nevertheless fall short as a fetus-incubator. The two sets of abilities don't have to go together.

And the environment after birth? What about the subtle effects of differences in mother's milk, or in the water used to dilute the baby's formula? What are the psychological factors of interaction with mother, with father, with foster parents, with siblings, with playmates? What are the sociological factors of the availability of books or of competitive games?

We don't know the answers to any of these things.

Nevertheless, there are people who have studied IQs of parents and children and are convinced there is a strong hereditary factor and that by following eugenic principles (mating "superior" parents) we can develop a "superior" strain of human beings.

One of those who apparently believes this is William B. Shockley, who is a Nobel Prize winner, and who admits to having contributed to the sperm bank. He is now seventy years old. Whether the sperm he is producing now are in quite the spanking condition they were a half-century ago, we can't be sure, of course.

And yet there are also bright people with not-so-bright kids; and vice versa.

For that matter take me. There is a general impression around that I'm bright. I've published, at the moment, 216 books, some fiction, some non-fiction, some for children, some for adults, some on each of the branches of science, some on literature, some on history, some on humor, and so on. Besides, I'm one of the best-paid lecturers in the country, and all my talks are off-the-cuff, whatever the fee. And on top of that, I'm Honorary Vice-President of Mensa, the high-IQ organization. This tends to make me seem highly intelligent and I must admit I do nothing to discourage the notion.

Well, then, suppose you wanted to produce me? What parents would you pick?

You would never pick mine, I'm quite sure. They were nice, warm, loving parents, who were reasonably intelligent – but only reasonably.

It was pure luck. The particular gene combination worked in my case, together with whatever it was that happened to me after birth.

Of course intelligent people might have intelligent youngsters a little oftener than not-so-intelligent parents might, but the chances are not great enough to make the eugenic experiment really worthwhile, in my opinion.

In fact, it might conceivably do harm. Suppose some scores of children are produced who, on the assumption that they are superior, are then treated with all possible tender and loving educational care. They are warmed before the benevolent fire of books and intellectual games and stimulating conversation.

Naturally, they will come closer to their full potential of intelligence than if they were just thrown in with a bunch of children and teachers taken at random from the very common herd. They will then *seem* superior, and we will be well on the way toward developing a kind of racist philosophy, a kind of feeling that there are several subspecies of human beings that either exist or can be developed and that differ from each other unalterably in intelligence.

We would then go on to develop, or try to develop, or convince ourselves that we have *indeed* developed, a ruling class of pure-bred intellectuals; a middle-class of respectable but inferior people; and a lower class of unutterable cads and swine, fit for enslavement.

It would be a terrible world. Such worlds have existed in the past and we still struggle with the marks of injustice and cruelty they have left behind.

What should we do, then?

Ought we not to consider that all over the world human beings with the potential of high intelligence are being born at all times? It may just possibly be that the percentage of such births is higher in some places than in others and among some groups of human beings than among others – but that doesn't matter. The percentage isn't zero anywhere.

If, then, we tried to develop a society that made sure that pregnant mothers were well cared for and babies well nourished everywhere, if psychological and social surroundings were healthy everywhere, if we developed a system of education that encouraged intelligence everywhere and if we made no artificial distinctions of appearance, language, or ways of life — if, in short, we developed a sane and just society — we would find ourselves harvesting highly intelligent children all over the world, and it would be a crop worth more to the human species than anything else conceivable.

And we would gather millions for sure, while the sperm banks were gathering dozens, perhaps.

But how can we manage to develop so ideal and Utopian a society as to harvest intelligence?

Considering the benefits to be derived, might we not *try*? If we even accomplish just a little in the way of improvement, even that little would grant us a richer harvest than the sperm banks will give us. And the additional intelligent people that would thus enter our society might help us accomplish still more — and so on.

We might as well try. At this stage in history, there is just about nothing left to lose.

# 8

# That Old-Time Violence

Violence is as human as thumbs and it's been with us a long, long time. Here's Herodotus (about 445 B.C.) talking in his "History" about King Gyges of Lydia: "As soon as Gyges was king he made an inroad on Miletus and Smyrna, and took the city of Colophon. Afterwards, however, though he reigned eight and thirty years, he did not perform a single noble exploit. I shall therefore make no further mention of him—"

Herodotus was certainly not going to waste time on an ignoble thing like a prosperous and peaceful 38-year reign. It was war and violence that drew the audience.

Or read Homer (about 800 B.C.), and in his *Iliad* you will have careful descriptions of exactly where the spear went in and where it came out. The great Achilles was given a choice between a long life spent ignobly (that is, as a hardworking farmer or herdsman, helping to feed humanity) or a short life full of immortal glory (as a wholesale slaughterer of those weaker than himself). Achilles chose the glory, and those who tell the tale expect us to admire Achilles and apparently we all do.

There's nothing new about violence; we've just used other words for it. We read about "glory," "gallantry," "derring-do," "knightly deeds," "noble exploits," "patriotic bravery," and it all comes down to violence whatever combination of glittering sounds you use.

And why? There may be many psychological reasons for it, but we can skip those. The fact is that the tales of violence served a practical, and even an essential, purpose.

Mankind lived by violence for uncounted thousands of years before history began. There were long ages in which human beings had to (if they could) kill animals for food, sometimes large animals who resented the attempt and resisted.

37

The mastodons and mammoths were driven to extinction by puny people swarming after them with nothing more deadly than hand-thrown spears. Would anyone run the risk of angering tons of bone and muscle if he weren't fortified with violent tales of great huntsmen meeting glorious deaths against ravening beasts.

There came a time when human beings won out over the animal kingdom once and for all. Thanks to mankind's possession of fire and of long-range missile weapons like the bow and arrow, no animal, not even the strongest and most deadly predator, could stand against the human onslaught.

That, however, didn't end the need for tales of violence, for having run out of other species to pit brain and weapon against, human beings fought other human beings. The skill and emotional drive developed by human beings over long ages of pitting feeble strength against tusks, fangs, claws, horns, and overreaching power, was now expended in a civil war to the death, one that has continued to this day.

Through most of history, when a city was taken its inhabitants were quite likely to be killed or enslaved. The women were first raped, of course, and then killed or enslaved. Even the Bible recommends mass slaughter as a routine matter of warfare. (See Deuteronomy 20:15-17, for instance.)

Under such circumstances, it was important to fight to the death, since one was in any case going to die, or worse, if one lost. So the people, or, at the very least, the warrior-class, were constantly fed tales of violence; of heroes who fought against overwhelming odds; of Hector standing against Achilles when all his fellow Trojans had fled; of Roland fighting off hordes of Saracens and disdaining to call for help; of the Knights of the Round Table taking on all challenges.

Youngsters had to get used to violence, had to have their hearts and minds hardened to it. They had to be made to feel the glory of fighting against odds and how sweet it was to give one's life for one's tribe, or one's city, or one's nation, or one's king, or one's fatherland, or one's motherland, or one's faith, or whatever other sounds are appropriate.

And now it's over! That old-time violence that's got us in its spell must stop!

Not because human beings have become good and sweet and gentle! Not because television makes violence too immediate by adding pictures to the sound!

Not at all! We've got to get rid of violence for the simple reason that it serves no purpose anymore, but points us all in a useless direction. It would appear that human enemies are no longer the prime threat to world survival.

The new enemies we have today — overpopulation, famine, pollution, scarcity — cannot be fought by violence. There is no way to crush those enemies, or slash them, or blast them, or vaporize them.

If they are to be defeated at all in their present incarnation, which threatens the whole world and all of mankind, rather than merely this tribe or that region, it must be by human cooperation and global determination. It is *that* which we had better start practicing. It is with tales of brotherhood and cooperation that we had better propagandize our children.

If we choose not to and if we continue to amuse ourselves with violence just because that worked for thousands of years, then the enemies that can't be conquered by violence will conquer *us*—and it will all be over.

# 9

# Little Green Men or Not?

When most people think about flying saucers or, as they are more austerely called, "unidentified flying objects" (UFOs), they think of spaceships coming from outside Earth and manned by extraterrestrial intelligences.

Is there any chance of this? Do the "little green men" really exist? There are arguments both for and against, pro and con.

*Pro.* There is, according to the best astronomical thinking today, a strong chance that life is very common in the universe. Our own galaxy is only one of perhaps a hundred billion, and our single galaxy has over a hundred billion stars in it.

Current theories about how stars are formed make it seem likely that planets are formed also, so that every star may have planets about it. Surely some of those planets would be rather like our Earth in chemistry and temperature.

Current theories about how life got its start make it seem that any planet with something like Earth's chemistry and temperature would be sure to develop life. One reasonable estimate advanced by an astronomer was that there might be as many as 640 million planets in our galaxy alone that are Earthlike and that bear life.

But on how many of these planets is there *intelligent* life? We can't say, but suppose that only one out of a million life-bearing planets develop intelligent life-forms and that only one out of ten of these develop a technological civilization more advanced than our own. There might still be as many as one hundred different advanced civilizations in our galaxy, and perhaps a hundred more in every other galaxy. Why shouldn't some of them have reached us?

*Con.* Assuming there are one hundred advanced intelligences in our own galaxy and that they are evenly spread throughout the galaxy, the nearest one would be about 10,000 light-years away. To cover that distance by any means we know of would take at least 10,000 years and very likely *much* longer. Why should anyone want to make such long journeys just to poke around curiously?

*Pro.* It is wrong to try to estimate the abilities of a far-distant advanced civilization, or their motives either. For one thing, the situation may not be average. The nearest advanced civilization may just happen to be only 100 light-years away, rather than 10,000.

Furthermore, because *we* know of no practical way of traveling faster than light doesn't mean an advanced civilization may not know of one. To an advanced civilization a distance of 100 light-years, or even 10,000 light-years, may be very little. They may be delighted to explore over long distances, just for the sake of exploring.

*Con.* But even if that were the case, it would make no sense to send so many spaceships so often (if we are to judge by the number of UFO reports over recent years). Surely we are not *that* interesting.

And if we *are* interesting, why not land and greet us? Or at least communicate with us without landing. They can't be afraid of us, since if they have advanced so far beyond us they can surely defend themselves against any puny threats we can offer.

On the other hand, if they want to be merely observers and don't want to interfere with the development of our civilization in any way, they should surely so handle their observations that we would not be continually aware of them.

*Pro.* Again, we can't try to guess what the motives of these explorers might be. What might seem logical to us might not seem so logical to them. They may not care if we see them, and they also may not care to say hello. Besides there are many reports of people who *have* seen the ships and have even been aboard. Surely some of these reports must have something to them.

*Con.* Eyewitness reports of actual spaceships and actual extraterrestrials are, in themselves, totally unreliable. There have been innumerable eyewitness reports of almost everything that most rational people do not care to accept—of ghosts, angels, levitation, zombies, werewolves, and so on.

What we really want, in this case, is something *material*; some object or artifact that is clearly not of human manufacture or earthly origin. These people who claim to have seen a spaceship or to have been inside one never end up with any button, rag, sheet of paper, or any other *object* that would substantiate their story.

*Pro.* But how else can you account for all the UFO reports? Even after you exclude reports that are incomplete or mistaken, that are gags or hoaxes, there still remain a large number of sightings that can't be explained by scientists within the present limits of knowledge. Aren't we forced to suppose these sightings are extraterrestrial spaceships?

*Con.* No, because we have no honest way of saying that the extraterrestrial spaceship is the *only* remaining explanation. If we can't think of any other, that may simply be because of a defect in our imagination or in our knowledge. To seize the easiest or most dramatic explanation as the only one left would be foolish. If an answer is unknown, then it is simply unknown. An Unidentified Flying Object is — unidentified.

The most serious and level-headed investigator of UFOs I know is J. Allen Hynek, a logical astronomer who is convinced that the UFO reports (or some of them, at least) are well worth serious investigation. He doesn't think that they represent extraterrestrial spaceships but he does suggest that they represent phenomena that lie outside the present structure of science and that understanding them will help us expand our knowledge and build a greatly enlarged structure of science.

He even thinks that the advance brought about by solving the UFO riddle could be so enormous that it would represent a "quantum jump" in some totally unexpected direction.

Well, perhaps; but that is only what he *believes*. He has no serious evidence to back his belief. The trouble is that, whatever the UFO phenomenon is, it comes and goes unexpectedly. There is no way of examining it systematically. It occasionally impinges on some of us and, more or less accidentally, is partially seen and then more or less inaccurately reported. We remain dependent on nothing more than occasional anecdotal accounts.

Dr. Hynek, after a quarter of a century of devoted and honest research, so far has ended with nothing. He not only has no solution, but he has no real idea of any possible solution. He only has his belief that when the solution comes it will be important.

He may be right, but there are at least equal grounds for believing that the solution may never come or that, when it comes, it will be unimportant.

# 10

# Don't You Believe?

One of the curses of being a well-known science-fiction writer is that unsophisticated people assume you to be soft in the head. They come to you for refuge from a hard and skeptical world.

Don't you believe in flying saucers, they ask me? Don't you believe in telepathy? — in ancient astronauts? — in the Bermuda triangle? — in life after death?

No, I reply. No, no, no, no, and again no.

One person recently, goaded into desperation by the litany of unrelieved negation, burst out, "Don't you believe in *anything*?"

"Yes," I said. "I believe evidence. I believe observation, measurement, and reasoning, confirmed by independent observers. I'll believe anything, no matter how wild and ridiculous, if there is evidence for it. The wilder and more ridiculous something is, however, the firmer and more solid the evidence will have to be."

For instance, where do I stand on telepathy, which I consider among the less wild suggestions along the fringes of knowledge?

I don't consider telepathy to be intrinsically impossible. After all, the brain produces a small electromagnetic field and the intensity of it wavers, rising and falling in irregular fashion, but with noticeable periodicities. These "brain waves" can be, and are, observed and measured by the technique of encephalography.

To be sure, the brain-waves are the overall product of some ten billion neurons, so that trying to make sense of them is like trying to make sense of the noise of the world's population all talking at once in all their various languages.

In listening to the world's overall human noise, we could tell the subsidence into a soft, drowsy hum when night covers a region; or the rise into

43

loud discordance at the coming of catastrophe. In the case of encephalog-raphy, there are changes from waking to sleeping, and vice versa, that can be detected. One can also detect the presence of a tumor or an epileptic seizure.

But we want something better than that; we want something that would be analogous to hearing the world's noise and picking out an individual conversation.

Might not specific thoughts affect the brain-wave pattern? Might not the wavering electromagnetic field then impress itself upon a neighbor-brain and induce that same thought upon it.

It is conceivable that this *might* happen, but the question is, is it conceiv-able that it *does* happen? Can one person detect another person's thoughts in actual practice?

Of course we can read thoughts indirectly. From the tone of a person's voice, from the expression of a person's face, from bits of a person's uncon-scious behavior, we can sometimes tell if that person is lying. We might even be able to make a shrewd guess as to what he (or she) is thinking. The more experienced we are, and the better we know the person we are studying, the more likely we are to guess his thoughts.

But that is not what we mean by telepathy. Can one person sense another's thoughts *directly*?

Well, consider—

If you were born with the ability to sense the thoughts of others, surely that would give you a considerable advantage. To sense what others don't realize is being sensed, to have advance warnings of others' intentions, to find that no secret is hidden—surely that would increase your security no end.

It seems to me, then, that telepathic ability has great survival value, and that even a very limited and rudimentary telepathic ability would have con-siderable survival value. Telepaths would be better off, would live longer, and would have more children (who would also be telepathic, most likely). The principles of natural selection, it seems to me, would surely see to it that more and more people would be more and more efficiently telepathic as time went on.

In fact, we might liken telepathy to vision. The ability to sense light and analyze it for information about one's surroundings offers such an advan-tage that almost all life-forms, even quite primitive ones, have eyes of one sort or another. Very efficient eyes long antedate humanity itself.

Therefore, the mere fact that we are now trying to find out if telepathy exists, that there is any question of it at all, is, in itself, very strong evidence that it does *not* exist. If it did exist, it would by now be an overriding ability that we would all take for granted.

—But wait, I may be going too far. Animals that live out their lives in total darkness are not likely to have eyes. It may be that telepathy has never developed on Earth because there have never been brains on Earth suffi-ciently complex to produce brain-waves worth detection, or to receive them, once produced. Only now, in the case of *Homo sapiens,* are the conditions right, and that just barely. Therefore, we are only now *beginning* to develop telepathy, so that very primitive effects are sometimes barely detectable in some people.

I find that hard to accept. Even simple brains have thoughts that could be powerful and worth receiving. The predator sneaking up on his prey must be thinking, at the very least, the equivalent of "−food−food−food−."

If the prey sees, hears, or smells the approaching predator, it is off at once, but surely that is not enough. The predator may be hidden, noiseless, and moving upwind. Would it not be useful for the prey to detect that "−food−" pulsing in the other brain?

I see a value to telepathy, an overriding survival value, that should have developed the ability in organisms with brains far too simple to develop complex ideas. We might as well argue that all animals but human beings should be deaf, since none of them have brains complex enough to be able to talk; or blind, since none of them have brains complex enough to be able to read. Hearing and sight have other, and more fundamental, functions than speaking and reading, and telepathy might well have other, and more fundamental, functions than carrying on an abstract conversation.

But maybe I'm wrong. Perhaps telepathy simply requires a more complex brain than sight and hearing do, and not all the need in the universe will force it into existence until the brain reaches a certain pitch of development. That would be why we're just beginning to detect it in a few quite rudimentary cases.

If that is so, doesn't it make sense to suppose that it is likely to show up in people with particularly efficient and complex brains? Yes, I know there are "idiot savants" who can do amazing things, but if telepathy can develop in backward brains, we're instantly back to wondering why it didn't develop in the lower animals.

If telepathy requires advanced brains, it will show up in particularly intelligent, shrewd, forceful, charismatic individuals, it seems to me. What's more, it would surely give them, even if it is present in only rudimentary form, a powerful advantage over others.

Might it not be, then, that telepathic powers explain how the leaders in politics, business, religion, science, and so on, come to be leaders? Might it not be just the touch of telepathy that does it?

I might believe that were it not that the world's leaders in every field have always shown a perfectly human capacity to be fooled, deceived, and betrayed. Julius Caesar clearly didn't know what was in the mind of Brutus. Napoleon I surely did not suspect his foreign minister, Talleyrand, to be playing the role of double agent for years. Hitler certainly didn't suspect that a bomb had been planted a few feet from him on July 20, 1944.

In other words, whether we consider the situation from the standpoint of biology or history, we see a world that simply doesn't make sense if telepathy exists.

I therefore conclude that the odds are enormously against the existence of telepathy.

In order to make me believe that telepathy exists, despite the evidence of the world around me, I would need very strong evidence, together with fool-proof reasoning, and this simply doesn't exist.

All that the proponents of telepathy can offer are anecdotal evidence and the kind of statistical analysis of guessing games that J. B. Rhine used to present. In these things, the possibilities of lies, hoaxes, or just honest distortion and wishful thinking are great enough to reduce it all to worthlessness in the face of the overwhelming evidence of the world we experience.

This is not to say that telepathy may not be possible sometime in the future. Conceivably, something of the sort may yet evolve as brains become still more complex.

Much more likely, in my opinion, is the chance that we may learn how to amplify, analyze, and interpret brain-waves to the point where we can "read minds" by instrument. I can even imagine people having combination amplifier/analyzers strapped unobtrusively behind the ear with fine leads attached to appropriate places on the skull, so that each person can broadcast his own thoughts and read those of others.

This, however, would be high technology and would not be the kind of telepathy that unsophisticated people ask me to "believe" in.

# 11

# Open Mind?

It is fashionable these days to accuse scientists of being dogmatic. If a scientist expresses confidence in some scientific conclusion, and acts as though opposition to that viewpoint is simply wrong, the opposition promptly denounces him for not having an open mind.

The opposition, of course, is fanatically convinced of the truth of *their* views and will not, under any circumstances, modify them in the slightest, but they are not scientists, you see, and they are not compelled to have open minds.

The result is that many scientists hesitate to attack the various kinds of nonsense that flood American society today, for fear of putting themselves in a bad light and of appearing dogmatic and close-minded. They therefore tend to keep quiet in the face of astrological fancies, pyramid fairytales, Bermuda Triangle myths, UFO mania, Velikovskian fables, creationist lunacy, and all the rest.

Since I am among those scientists who attack nonsense without hesitation, and as strongly as I can, I am sometimes accused of "overreacting," and of "overstating" my case. My usual response to these fainthearts is to ask whether they have the guts to say the same to the opposition.

I don't for one minute expect that my defense of rationalism is going to make any difference to the many unsophisticated people who enjoy believing the nonsense they read and hear, and who have no way of separating folly from sense, but I do have my own self-respect to consider. However hopeless the fight, I cannot simply surrender.

See here! The earth is not flat!

That is a scientific conclusion, based on careful observation and reasoning, and that conclusion is older than Aristotle. Ever since his time, enormous quantities of additional information have been wrung out of the universe, and *all of it* supports the conclusion that the earth is not flat and is, indeed, spherical.

To be sure, careful observations have refined the conclusion. The earth is not a perfect mathematical sphere. Because of its rotation, it is an oblate spheroid, but to such a slight degree that the difference from perfect sphericity would be imperceptible if we viewed the earth from space with the unaided eye.

Nor is it a perfect oblate spheroid. Satellite observations have determined very slight departures from that and, of course, there is the lumpiness of hills and valleys.

However, the small deviations from perfect sphericity do not force me to conclude that, therefore, the earth is flat.

Yet there are people who believe the earth is flat. I don't mean just primitive tribesmen who accept the hasty evidence of their eyes without consideration. I mean presumably educated Americans and Europeans who argue, with apparent sincerity, that all the evidence cited for nonflatness is either wrong or misunderstood, and that the observations of the astronauts, by eye and by camera, have been "faked."

How shall I treat these people? With respect? Shall I offer to compromise? Shall I say, "Well, scientific open-mindedness compels me to agree that the earth *may* be flat, or is at least partly flat"? Is that the only way I can avoid "overreacting," and "overstating my case"?

Never!

And that goes for all other varieties of nonsense. If I think that certain views are crackpot, I intend to say so.

# 12

# The Role of the Heretic

What does one do with a heretic?

We know the answer if the "one" referred to is a powerful religious orthodoxy. The heretic can be burned at the stake.

If the "one" is a powerful political orthodoxy, the heretic can be sent to a concentration camp.

If the "one" is a powerful socioeconomic orthodoxy, the heretic can be prevented from earning a living.

But what if the "one" is a powerful scientific orthodoxy?

In that case, very little can be done, because even the most powerful scientific orthodoxy is not very powerful. To be sure, if the heretic is himself a scientist and depends on some organized scientific pursuit for his living or for his renown, things can be made hard for him. He can be deprived of government grants, of prestige-filled appointments, of access to the learned journals.

This is bad enough, to be sure, and not lightly to be condoned, but it is peanuts compared to the punishments that could be, and sometimes are, visited on heretics by the other orthodoxies.

Then, too, the religious, political, and socioeconomic orthodoxies are universal in their power. A religious orthodoxy in full flight visits its punishments not on priests alone; nor a political one on politicians alone; nor a socioeconomic one on society's leaders alone. No one is immune to their displeasure.

The scientific orthodoxy, however, is completely helpless if the heretic is not himself a professional scientist—if he does not depend on grants or appointments, and if he places his views before the world through some medium other than the learned journals.

Therefore, if we are to consider scientific heretics, we must understand that there are two varieties with different powers and different immunities.

Let's consider the two kinds of scientific heretics.

1. There are those who arise from within the professional world of science and who are subject to punishment by the orthodoxy. We might call these heretics from within "endoheretics."

2. There are those who arise from outside the professional world of science and who are immune to direct punishment by the orthodoxy. These heretics from without are the "exoheretics."

Of the two, the endoheretics are far less well known to the general public. The endoheretic speaks in the same language as does the orthodoxy, and both views, the endoheretical and the orthodox, are equally obscure to the nonscientist, who can, generally speaking, understand neither the one nor the other nor the nature of the conflict between them.

It follows that, if we consider the great endoheresies of the past, we find that the general public is not ordinarily involved. In the few cases where they are involved, it is almost invariably on the side of orthodoxy.

The patron saint of all scientific heresies, Galileo, was, of course, an endoheretic. He was as deeply versed in Aristotelian physics and Ptolemaic astronomy, which he dethroned, as were any of his Aristotelian/Ptolemaic opponents.

And, since in those days and in his particular society, the scientific and religious orthodoxies were the same, Galileo had to run far greater risks than later endoheretics did. Facing the Inquisition, he had to consider the possibility not of a canceled grant but of physical torture.

Yet we cannot suppose that there was any great popular outcry on behalf of the rebel. The general public was not concerned or even aware of the dispute. Had they been made aware, they would have certainly sided with orthodoxy.

Next to Galileo, the greatest of the endoheretics was Charles Darwin, whose views on the evolution of species through the blind action of chance variation and natural selection turned biology upside-down. Here the general public *did* know of the controversy and *did,* in a very general and rough way, have a dim view of what it was about. And the public was definitely on the side of the orthodoxy.

The public has remained anti-evolution to this day. Science has accepted Darwin without, at this time, respectable dissent. The more sophisticated churches no longer quarrel publicly with it. But the general public, in what is probably the majority opinion if a vote were to be taken, stubbornly adheres to the tenets of a lost and dead orthodoxy of a century and a quarter ago.

Galileo and Darwin won out. A number of the endoheretics did. Never by public pressure, however. Never by majority votes of the general public.

They won out because science is a self-correcting structure and because observation, experimentation, and reasoning eventually support those heresies that represent a more accurate view of the universe and bury those orthodoxies that are outpaced.

In the process, orthodoxy gets a bad press. Looking back on the history of science, it would seem that every endoheretic was right, that each wore the white hat of heroism against an evil and short-sighted orthodoxy.

But that is only because the history of science is naturally selective. Only the endoheretic who was, in the end, shown to be right, makes his mark. For each of those, there may have been, let us say, fifty endoheretics who were quite wrong, and whose views are therefore scarcely remembered, and who are not recorded even as a footnote in the history books — or, if they are, it is for other nonheretical work.

What, then, would you have the orthodoxy do? Is it better to reject everything and be wrong once in fifty times — or accept everything and be wrong forty-nine in fifty times and send science down endless blind alleys as a result?

The best, of course, would be neither. The best would be to reject the forty-nine wrong out of hand and to accept and cherish the one right.

Unfortunately, the day that the endoheretical pearl shines out so obviously amid the endoheretical garbage as to be easily plucked out is the day of the millennium. There is, alas, no easy way of distinguishing the stroke of intuitional genius from the stroke of folly. In fact, there has been many an utterly nonsensical suggestion that has seemed to carry much more of the mark of truth than the cleverly insightful stroke of genius.

There is no way, then, of dealing with the endoheresies other than by a firm (but not blind or spiteful) opposition. Each must run the gauntlet that alone can test it.

And it works. There is delay and heartbreak often enough, but it works. However grimly and slowly the self-correcting process of science proceeds, that the process exists at all is a matter of pride to scientists. Science remains the only one of man's intellectual endeavors that is self-correcting at all.

The problem of endoheresy, then, is not a truly serious one for science (though it may be, we all know, for the individual endoheretic) and is not one that must be ironed out in public.

But what of exoheresy?

We had better first be sure of what we mean by an exoheretic. Science is split into endless specialties, and a specialist who is narrow-minded and insecure may see as an "outsider" anyone who is not bull's-eye on target within the specialty.

Robert Mayer was a physician and James P. Joule was a brewer who dabbled in physics. Neither had academic credentials, and the fact that both saw the existence of the law of conservation of energy went for nothing. Neither could get his views accepted. Hermann Helmholtz, third in line, was a full academician, and he gets the credit.

When Jacobus van't Hoff worked out the scheme of the tretravalent carbon atom, the orthodox chemist Adolf Kolbe denounced the new concept intemperately, specifically, and contemptuously, mentioning the fact that van't Hoff was teaching at a veterinary school.

But we can't go along with this. If we wish to be fine enough and narrow enough, then all scientific heretics are exoheretics in the eyes of the sufficiently orthodox and the term becomes meaningless.

Nor should we label as exoheretics those who are not formally educated but who, through self-education, have reached the pitch of professional excellence.

Let us, instead, understand the word exoheretic to refer only to someone who is a real outsider, one who does not understand the painstaking structure built up by science, and who therefore must attack it without understanding.

The typical exoheretic is so unaware of the intimate structure of science, of the methods and philosophy of science, of the very language of science, that his views are virtually unintelligible from the scientific standpoint. As a consequence, he is generally ignored by scientists. If exoheretical views are forced upon scientists, the reaction is bound to be puzzlement or amusement or contempt. In any case, it would be exceptional if the exoheresy were deemed worthy of any sort of comment.

In frustration, the exoheretic is then very likely to appeal over the heads of the scientists to the general public. He may even do it successfully, since his inability to speak the language of science does not necessarily prevent him from speaking the language of the public.

The appeal to the public is, of course, valueless from the scientific standpoint. The findings of science, after all, cannot be canceled or reversed by majority vote, or, for that matter, by the highest legislative or executive fiat. If every government in the world declared, officially, that the earth was flat, and if every scientist were forbidden to argue the contrary, the earth would nevertheless remain spheroidal and every scrap of evidence maintaining that conclusion would still exist.

Nevertheless, the appeal to the public has other rewards than that of establishing scientific proof.

1. If the public responds favorably, it is soul-satisfying. The exoheretic can easily convince himself that the fact that he is the center of a cult demonstrates the value of his views. He can easily argue himself into believing that people would not flock to nonsense, though all history shows otherwise.

2. If the public responds favorably, the results can be lucrative. It is well known that books and lectures dealing favorably with a popular cult do far better than do books and lectures debunking it. This is not altered by the

fact that the books in favor may be poorly written and reasoned, whereas the books against may be models of lucidity and rationality.

3. If the public responds favorably, scientists may be hounded into open opposition and may express, with injudicious force, their opinion of the obvious nonsense of the exoheretical views. This very opposition, casting the exoheretic into the role of martyr, works to accentuate the first two advantages.

The fact remains, nevertheless, public support or not, that the exoheretic virtually never proves to be right. (How can he, when he, quite literally, doesn't know what he's talking about.) Of course, he may prove to have said something somewhere in his flood of words that bears some resemblance to something that later proves to be so, and this coincidental concurrence of word and fact may be hailed as proving all the rest of the corpus of his work. This, however, has only cultic value.

We see, then, the vast difference between endoheretics and exoheretics:

1a. The public is, generally, not interested in the endoheretic; or, if aware of him at all, is hostile to him. The endoheretic therefore rarely profits from his heresy in any material way.*

1b. The public, on the other hand, can be very interested in the exoheretic and can support him with a partisan and even religious fervor, so that the exoheretic may, in a material way, profit very considerably by his heresy.

2a. The endoheretic is sometimes right, and, indeed, since startling scientific advances usually begin as heresies, some of the greatest names in science have been endoheretics.

2b. The exoheretic, on the other hand, is virtually never right, and the history of science contains no great advance, to my knowledge, initiated by an exoheretic.

One might combine these generalizations and, working backward (not always a safe procedure), state that when a view denounced by scientists as false is, nevertheless, popular with the general public, the mere fact of that popularity is strong evidence in favor of its worthlessness.

It is on the basis of public popularity of particular beliefs, for instance, that I, even without personal investigation of such matters, feel it safe to be extremely skeptical about ancient astronauts, or about modern astronauts in UFOs, or about the value of talking to plants, or about psionic phenomena, or about spiritualism, or about astrology.†

---

* I must qualify these generalizations because there are exceptions of course. Edward Jenner, who advanced the endoheretical technique of smallpox vaccination, was accepted eagerly by the public, and profited materially as a result.

† Of course, I would also have used this line of reasoning to feel it safe to be skeptical about the value of smallpox vaccination, but the facts would have converted me within a year in that case.

And this brings me to Velikovskianism at last.

Of all the exoheretics, Velikovsky has come closest to discomfiting the science he has attacked, and has most succesfully forced himself to be taken seriously. Why is that? Well—

1. Velikovsky was a psychiatrist and so he had training in a scientific specialty of sorts and was not an utter exoheretic. What's more, he had the faculty of sounding as though he knew what he was talking about when he invaded the precincts of astronomy. He didn't make very many elementary mistakes and he was able to use the language of science sufficiently well to impress a layman, at any rate.

2. He was an interesting writer. It's fun to read his books. I myself have read every book he published. Although he didn't lure me into accepting his views, I can well see how those less knowledgeable in the fields Velikovsky dealt with succumbed.

3. Velikovsky's views in *Worlds in Collision* are designed to demonstrate that the Bible has a great deal of literal truth in it; that the miraculous events described in the Old Testament really happened as described. To be sure, Velikovsky abandons the hypothesis that divine intervention caused the miracles and substitutes a far less emotionally satisfactory hypothesis involving planetary ping-pong, but that scarcely alters the fact that, in our theistic society, any claimed finding that tends to demonstrate the truth of the Bible is highly likely to meet with general favor.

These three points are enough in themselves to explain Velikovsky's popularity. Supply the public with something that is amusing, that sounds scholarly, and that supports something it wants to believe, and surely you need nothing more. Erich von Däniken and his inane theories of ancient astronauts have proven successful on nothing more than this, even though his books are less amusing than Velikovsky's, sound less scholarly, and support something less foolproof than belief in the Bible.

It was the good fortune of Velikovsky, however, to go beyond this. Because of the climate of the times when *Worlds in Collision* was published, there was an astronomical overreaction. The initial appearance of his views in *Harper's* and the anxiety of the editorial staff of that magazine to achieve an improved circulation by inflating the significance of the article, goaded some astronomers into an attempt at censorship. To paraphrase Fouchet, this was worse than immoral; it was a blunder. Velikovsky lived on it for a quarter of a century.

The fact that Velikovsky could portray himself as a persecuted martyr cast a Galilean glow upon all his endeavors, and canceled out any attempt on the part of astronomers to demonstrate, clearly and dispassionately, the errors in the Velikovskian view. All attempts in this direction could be (and were) dismissed as persecution.

It also gives a glow of heroism to Velikovsky's followers. They can attack an orthodoxy, something that is ordinarily accepted as a courageous thing to do, and yet can do so with complete safety since in actual fact (as opposed to Velikovskian fantasy) the orthodoxy does not, and indeed cannot, strike back.

From the standpoint of science, then, is Velikovskianism nothing but an irritation and a waste of time?

Not at all. It has its enormous benefits.

For one thing, Velikovskianism, and indeed, any exoheretical view that becomes prominent enough to force its views on science, acts to puncture scientific complacency—and that is good.

An exoheresy may cause scientists to bestir themselves for the purpose of reexamining the bases of their beliefs, even if only to gather firm and logical reasons for the rejection of the exoheresy—and that is good, too.

An exoheresy may cause scientific activity that, in serendipitous fashion, may uncover something worthwhile that has nothing to do with the exoheresy—and that would be very good, if it happens.

I hope scientific orthodoxies never remain unchallenged. Science is in far greater danger from an absence of challenge than from the coming of any number of even absurd challenges. Science, unchallenged, can become arthritic and senile, whereas the most absurd challenge may help to stir the blood and tone the muscles of the body of science.

Therefore, it was altogether fitting and proper that Velikovsky was given a hearing at the AAAS meetings. Though one could be sure from the start that nothing scientists would say would in the least move the Velikovskians and that no amount of mere logic would shake their faith, it was still a good thing—for science.

And there is this comfort, too. An exoheresy that is patently in error cannot change the universe to conform to itself. However popular it may be and however irritatingly it may survive refutation, its falseness condemns it—in the end—to nothingness.

# Part III

# Population

# 13

# The Good Earth Is Dying

How many people is the earth able to sustain?

The question is incomplete as it stands. One must modify the question by asking further: At what level of technology? And modify it still further by asking: At what level of human dignity?

As for technology, perhaps we can simply ask for the best. We can say that the more advanced technology is, the more people the earth can support, so let us not stint. After all, technology could give us the atomic bomb and put men on the moon and we should set no limits to it.

Let us accept, then, the dream that technology is infinitely capable and proceed from there. How many people can the earth sustain assuming that technology can solve all reasonable problems?

To begin with, it is estimated that there are twenty million million tons of living tissue on the earth, of which 10 percent, or two million million tons, is animal life. As a first approximation, this may be considered a maximum, since plant life cannot increase in mass without an increase in solar radiation or an increase in its own efficiency in handling sunlight. Animal life cannot increase in mass without an increase in the plant mass that serves as its ultimate food.

The mass of humanity has been increasing throughout history; and it is still increasing, but is doing so at the expense of other forms of animal life. Every additional kilogram of humanity has meant, as a matter of absolute necessity, one less kilogram of nonhuman animal life. We might argue, then, that the earth can support, as a maximum, a mass of mankind equal to the present mass of all animal life. At that point, the number of human beings on the earth would be forty million million, or over eleven thousand times the present number. And no other species of animal life would then exist.

What will this mean? The total surface of the earth is five hundred twenty million square kilometers, so that when human population reaches its ultimate number, the average density of population will be eighty thousand per square kilometer — twice the density of New York's island of Manhattan. Imagine such a density everywhere if the earth's population is spread out evenly — including over the polar regions, the deserts, and the oceans.

Can we imagine, then, a huge, world-girdling complex of high-rise apartments (over both land and sea) for housing, for offices, for industry? The roof of this complex will be given over entirely to plant growth; either algae, which are completely edible, or higher plants that must be treated appropriately to make all parts edible.

At frequent intervals, there will be conduits down which water and plant products will pour. The plant products will be filtered out, dried, treated, and prepared for food, while the water is returned to the tanks above. Other conduits, leading upward, will bring up the raw minerals needed for plant growth, consisting of (what else) human wastes and chopped-up human corpses. And at this point, of course, no further increase in human numbers is possible; so rigid population planning would then be necessary if it had not been before.

But if this number can be supported in theory, does it represent a kind of life — and this is for each of you to ponder — consonant with human dignity?

Can we buy space and time by transferring human beings to the moon? To Mars?

Consider — How long, under present conditions, will it take us to reach the global high-rise? At present, the earth's population is thirty-six hundred million and it is increasing at a rate that will double the figure in thirty-five years. Let us suppose that this rate of increase can be maintained. In that case, the ultimate population of forty million million will be reached in 465 years. The global high-rise will be in full splendor in A.D. 2436.

In that case, how many men do you think it will be possible to place, and support, on the moon, Mars, and elsewhere in the next 465 years? Be reasonable. Subtract your figure from forty million million and ask yourself if the contribution the other worlds can make is significant.

Can we buy further time by going beyond the sun? Can we make use of hydrogen fusion power to irradiate plant life? Or can we make food in the laboratory, with artificial systems and synthetic catalysts, and declare ourselves independent of the plant world altogether?

But that requires energy and here we come to another point. The sun pours down on the earth's day-side, some fifteen thousand times as much energy per day as mankind now uses. The earth's night-side must radiate exactly that much heat back into space if the earth's average temperature is to be maintained. If mankind adds to the heat on earth by burning coal, that additional energy must also be radiated out to space; and to accomplish this the earth's average temperature must rise slightly.

At present, man's addition to solar energy produces a terrestrial temperature-rise that is utterly insignificant; that addition, however, is doubling every twenty years. At this rate, in a hundred sixty-five years (by 2136) mankind's contribution to the heat that the earth must radiate away will amount to one percent of the sun's supply, and this will begin to produce unacceptable changes in the earth's temperature.

So, far from helping ourselves with further energy expenditures in the global high-rise world of A.D. 2436, we must accept a limitation of energy expenditure a full three centuries earlier, when man's population is less than a five-hundredth of its ultimate. We might improve matters by increasing the efficiency with which energy is used; but the efficiency cannot rise above a hundred percent, and that does *not* represent an enormous increase over present levels.

*But,* and this is a large "but," can we really depend on technology to make the necessary advances to bring us to energy-limits safely in a century and a half? By then the population, at the present level of increase, will be twenty times what it is today; and to bring man's level of nourishment to a desirable point, we would need a fortyfold increase in the food supply. We would also have to ask technology not only to arrange the necessary hundred-fifty-fold increase in energy utilization in a century and a half but to arrange to take care of what will be, very likely, a hundred-fifty-fold increase in environmental pollution and in waste production of all kinds.

How do things look at present?

Far from making strides to keep up with the population increase, technology is falling visibly behind. How can the global high-rise be a reasonable future vision when present-day housing is steadily deteriorating even in the most advanced nations? How can we reach our limit of energy expenditure when New York City finds itself, each year, with a growing deficit of power supply. Only yesterday, the third landing of men on the moon caused television viewing to go up, and a cutback in electrical voltage was immediately made necessary.

The earth's population will be at least six thousand million in the year 2000. Will the planet's technology be able to support that population even at present-day, wholly unsatisfactory levels? Will human dignity be compatible with such a population (let alone with forty million million), when in our cities *today* human dignity is disappearing; when it is impossible to walk safely by night (and often by day) in the largest city of the world's most technologically advanced nation.

Let us not look into the future at all, then. Let us gaze firmly at the present. The United States is the richest nation on earth and every other nation would like to be at least as rich. But the United States can live as it does only because it consumes slightly more than half of all the energy produced on earth for human consumption—although it has only a sixteenth of the earth's population.

What, then, if some wizard were to wave his magic wand and produce an earth on which every part of the population everywhere were able to live at the scale and the standard of Americans? In that case, the rate of energy expenditure would increase instantly to eight times the present world level and, inevitably, the production of waste and pollution would increase similarly—this with no increase in population at all.

And can present-day technology supply an eightfold increase in energy utilization (and that of other resources as well) and handle an eightfold increase in waste and pollution, when it is falling desperately short of supplying and handling present levels? Do you ask for time in which technology can arrange for such an eightfold increase? Very well, but in that time, population will increase, too, very likely more than eightfold.

In short, then, to the revised question, How many people is the earth able to sustain at a desirable level of technology and dignity? there can be only a short and horrifying answer—

Fewer than now exist!

The earth cannot support its present population at the average level of the American standard of living. Perhaps, at the moment, it can only support five hundred million people at that standard. Nor can technology improve itself to better this mark, with the present population clamoring for what it cannot have and with that population growing at a terrible rate.

What, then, will happen?

If matters continue as they are now going, there will be a continuing decline in the well-being of the individual human being on earth. Calories per mouth will decrease; available living quarters will dwindle; attainable comfort will diminish. What is more, in the increasing desperation to reverse this, man may well make wild attempts to race the technology-engine at all costs and will then further pollute the environment and decrease its ability to support mankind. With all this taking place, there will be a struggle of man against man, with each striving to grasp an adequate share of a shrinking life-potential. And there cannot help but be an intensification of the human-jungle characteristics of our centers of population.

In not too long a time, the population increase will halt; but for the worst possible reason—there will be a catastrophic rise in the death rate. The famines will come, the pestilence will strike, civil disorder will intensify, and by A.D. 2000 some governmental leader may well be desperate enough to push the nuclear button.

How to prevent this, then?

We must stop living by the code of the past. We have, over man's history, developed a way of life that fit an empty planet and a short existence marked by high infant-mortality and brief life-expectancy. In such a world there was a virtue in having many children, in striving for growth in numbers and power, in expansion into endless space, in total commitment to that limited portion of mankind that could make up part of a viable society.

But none of this is so any longer. At the moment, child mortality is low, life expectancy high, the earth full. There are no empty spaces of worth, and so interdependent is man that it is no longer safe to confine loyalty to only a portion of mankind.

What was common sense in a world that once existed has become myth in the totally different world that now exists, and suicidal myth at that.

In our overpopulated world we can no longer behave as though woman's only function in life is to be a baby-producing machine. We can no longer believe that the greatest blessing a man can have is many children.

Motherhood is a privilege we must literally ration, for children, if produced indiscriminately, will be the death of the human race; and any woman who deliberately has more than two children is committing a crime against humanity.

We also have to alter our attitude toward sex. Through all the history of man it has been necessary to have as many children as possible, and sex has been made the handmaiden of that fact. Men and women have been taught that the only function of sex is to have children; that otherwise it is a bestial and wicked act. Men and women have been taught that only those forms of sex that make conception possible are tolerable; that everything else is perverse, unnatural, and criminal.

Yet we can no longer indulge in such views. Since sex cannot be suppressed, it must be divorced from conception. Birth control must become the norm and sex must become a social and interpersonal act rather than a child-centered act.

We also have to alter our attitude toward growth. The feeling of "bigger and better" that bore up mankind through his millennia on this planet must be abandoned. We have reached the stage where bigger is no longer better. Although the notion of more people, more crops, more products, more machines, more gadgets—more, more, more—has worked, after a fashion, up to this generation, it will no longer work. If we attempt to force it to work, it will kill us rather quickly.

In our new and finite world, where for the first time in history we have reached, or are reaching, our limit, we must accept the fact of limit. We must limit our population, limit the strain we put on the earth's resources, limit the wastes we produce, limit the energy we use. We must preserve. We must preserve the environment, preserve the other forms of life that contribute to the fabric and viability of the biosphere, preserve beauty and comfort. And if we do limit and preserve, we will have room for deeper growth even so—growth in knowledge, in wisdom, and in love for one another.

We also have to alter our attitude toward localism. We can no longer expect to profit by another's misfortune. We can no longer settle quarrels by wholesale murder. The price has escalated to an unacceptable level. World War II was the last war that could be fought on this planet by major powers using maximum force. Since 1945, only limited wars have been conceivable, and even these have been monumental stupidities, as the situations in Southeast Asia and in the Middle East make clear.

The world is too small for the kind of localism that leads to wars. We can have special pride in our country, our language, and our literature, our customs and culture and tradition, but it has to be the abstract pride we have in our baseball team or our college—a pride that cannot and must not be backed by force of arms.

Localism doesn't even have the virtue of being useful in times of peace. The problems of the world today are planetary in scope. No one nation, not even if it is as rich as the United States, as centralized as the Soviet Union, or as populous as China, can solve its important problems today. No matter how a nation stabilizes population within its own borders, no matter how it rationalizes the use of its own resources, no matter how it conserves its own environment, all would come to nothing if the rest of the world continued its rabbit-multiplication and its poisoning of free will.

Even if every nation sincerely took measures, independent of each other, to correct the situation, the solutions one nation arrived at would not necessarily match those of its neighbors, and all might fail.

To put it bluntly, planetary problems require a planetary program and a planetary solution, and that means cooperation among nations, *real* cooperation. To put it still more bluntly, we need a world government that can come to logical and humane decisions and can then enforce them.

This does not mean a world government that will enforce conformity in every respect. The cultural diversity of mankind is surely a most valuable characteristic and it must be preserved—but not where it will threaten the species with suicide.

All these requirements for change go against the grain. Who really wants to downgrade motherhood and regard babies as enemies? Who is comfortable at the thought of dissociating sex and parenthood? Who is ready to submit his national pride to a truly effective world government? Who is willing to abandon the attempt to get as much as possible out of the world and settle instead for a controlled and limited exploitation?

Yet the logic of events is actually forcing us in that direction, willing or not. The birth rate is dropping in those nations that have access to birth-control methods. Sexual mores are loosening everywhere. The people are growing more concerned about the environment, and the clamor for cleaner air, water, and soil is becoming louder every day.

Most of all, and most heartening, localism is retreating. There is increasing social and economic cooperation among neighboring nations; a stronger drive in the direction of regionalism. More important still, there is a clear understanding that a major war, particularly one between the United States and the Soviet Union, is inadmissible. These two superpowers have quarreled at levels of intensity that at any time up to the 1930s would have meant war—and now those quarrels do not even bring about a rupture in diplomatic relations. Not only must these nations not fight; they must not even snub each other.

But this motion in the right direction does not seem to be a matter of choice. Rather, stubborn humanity is inching forward to help itself only because the pressure of circumstance has closed all other passageways.

And this motion in the right direction is not fast enough. The population increase continues to outpace the education for birth control; the environment continues to deteriorate more rapidly than we can bring ourselves to correct matters; and, worse, nations still stubbornly quarrel, and continue to place local pride over the life and death of the species.

We must not only reorient our thinking toward motherhood, sex, growth and localism as we are beginning to do; but we must do it more *quickly*. Our society cannot survive another generation of the steadily intensifying stresses placed on it. If we continue as we are and change no faster, then by 2000 the technological structure of human society will almost certainly have been destroyed. Mankind, having been reduced to barbarism, may possibly be on the way to extinction. The planet itself may find its ability to support life seriously compromised.

The good earth is dying; so in the name of humanity let us move. Let us make our hard but necessary decisions. Let us do it quickly. Let us do it now.

# 14

# The Price of Survival

The basic danger that civilization faces today is too many people. The earth's population at the moment is four billion. Each year there are eighty million more people than the year before.

Are you concerned about famines in the world? They come about because we're having trouble filling four billion mouths. If it's difficult to do so now, how about next year when there will be eighty million more, and the year after, when there will be another eighty million more? By 2010, at this rate, there will be *eight* billion people in the world and the population will be increasing by *a hundred sixty million* a year.

Are you bothered by inflation? By recession? If more and more people demand more and more goods out of a no longer increasing supply, those goods must be rationed. If not, there must be inflation, because that too is a kind of rationing—at the expense of the poor. And, if supplies go down while the cost of labor goes up, businesses must contract and unemployment rises.

Are you troubled by declining resources and deteriorating environment? If you try to provide goods and services for more and more people each year, the resources you need must be used up faster and faster, while the wastes produced pile up and disfigure and poison the environment.

Do you fear the danger of war? The increase of terrorism? The heightening of alienation? If you make it harder and harder for more and more people to get the goods they need for their bodies and the space they need for their souls, then you *must* have increasing alienation—and friction, and violence, and terrorism. Eventually, you may even have the pushing of the nuclear button.

It comes down to overpopulation, and whether by starvation or by nuclear holocaust, or even by just rattling to death, our industrial civilization will not survive. It may not even survive this century.

Can we shrug off our civilization and say that the world would be better off, anyway, if it comes down to shepherds watching their flocks and farmers sowing their fields by hand, in a happy pastoral spring? Unfortunately, we cannot.

A world of farming and herding cannot, at best, support more than a billion people on earth—perhaps considerably less in the world that will be left after we are through tearing it apart in the attempt to feed and supply our crowding billions. And what will we do with all those billions if the complex technology that supports them (however imperfectly) falls apart?

There is no way out. If we are to enter the twenty-first century with a reasonable hope of avoiding the greatest catastrophe in human history, it will be because our technology is still in operation. And that can only be because humanity has brought the population problem under control. We must be able to look forward to a twenty-first century in which the population has first leveled off and then begins to decrease to some level the earth can support decently and graciously.

I do not say that mankind can do this: perhaps it cannot. Nevertheless, unless this is done, civilization will not survive long past the turn of the century and the catastrophe will be appalling. Negative population growth by the opening decades of the next century is the price of survival.

And perhaps we *can* do it. Population growth is in the process of being checked in the United States and in some of the other advanced countries. (This in itself, of course, offers us no assurance that we can survive while the rest of the world collapses. As population rises in the world generally, the crunch in food, energy, and other resources will grow rapidly worse. And, since we depend on the rest of the world for many of those resources, we will experience that crunch, too.)

The danger of unrestrained growth is becoming more and more apparent to more and more of the world's nations, and it may be that the world generally will follow the American example and will pay this price of survival.

But, if it does, what will be required of the twenty-first century? What will the world be like in a time of negative population growth?

Consider, first, that there are only two ways of bringing about a negative population growth. We can raise the death rate, shortening the length of life. Or we can allow life to remain long and lower the birth rate.

In choosing between these two alternatives, consider what methods we might have to use in order to raise the death rate. Shall we cut down food and let more people die of starvation? Shall we spread disease? Shall we order wholesale executions? There are no humane, decent methods of allowing the death rate to go up, and civilization would not survive a world policy of selective murder. If mankind has a real choice, it must surely opt for a reduced birth rate.

As the price of survival, then, we must enter the twenty-first century with negative population growth in the process of being achieved through a sharp decline in the birth rate.

And what will a low-birth-rate world be like?

If we have difficulty trying to imagine such a situation, we might ask ourselves, first, what a *high*-birth-rate world would be like. That, at least, we have experienced.

Through most of man's history, a high birth rate has been necessary to survival. Without our present industrial civilization and without our present level of science and technology, the death rate was high. Starvation and famine were common; disease and plagues were endemic. It took all the babies that could be produced to make up for the hurricane of deaths that constantly plagued the human species.

It is, of course, women who must produce these babies, who must carry the fetuses for nine months, who must suckle the infants for a year or so after birth and then (held to them by the tight bonds of affection) care for them for additional years afterward. (Yes, the man plays a vital role, but that role takes only a few minutes per child.)

Any woman who must produce many children under pre-technological conditions has little time for anything else. Indeed, the price of survival under those conditions is to see to it that women are *not* given anything else to do; that they are kept firmly to their task of baby machine and domestic drudge. Only in that way can the number of babies produced be high enough to keep the population from dwindling away.

Since the task of baby machine and domestic drudge has limited attractions, women had to be propagandized into accepting the role. Probably the longest-sustained and most unrelenting propaganda campaign in history has been the one that has been used to persuade women that wife-and-motherhood was their total fulfillment, that there could be no nobler task than that of rocking the cradle, that any attempt to step out into the world outside would deprive them of something precious called "femininity."

The two go together. A high-birth-rate world means women's subservience. Without women's subservience we can't have a high-birth-rate world.

Well, then, what about a low-birth-rate world? In such a world will women be set free?

Yes, they will. It is not even a matter of choice. They *must* be set free. Consider—

In a world of multi-billions, where the population must be made to drop rather rapidly for the sake of survival, women must *not* have many babies. They must certainly not have more than two; and, if they want to have only one, or none at all, that is all right, too. (There is no danger of species-suicide. If at any time the population seems too low, it can, with very little trouble, be doubled in thirty-five years or even less.)

But surely it is not enough to say to women: "Don't have babies!" Having babies takes time both before and after birth, and not having them leaves a long gap in a woman's life that must be filled with something else. Then, too, having babies *does* have its positive aspects; the process *is* an enriching experience; babies *are* lovable. Not having them leaves a large hole in a woman's emotional needs that must be filled with something else.

If the gap is not filled, if the empty spaces are left empty, then undoubtedly women will want children and will strive to have them.

Can we then launch an even more intense and sustained propaganda campaign to convince women that an empty life is the highest ideal to which she can aspire? Can we tell her that a role as a man's shadow, doing for him whatever he does not feel like doing for himself, is enough to keep her happy?

It wouldn't be right or just to do so, but never mind that. The important thing is that it wouldn't work.

If, in the twenty-first century, we want to keep women out of the nursery, we have to substitute another kind of life. We can't replace something with nothing. We have to replace something with something.

And what is the replacement?

Through history, mankind has developed two general types of life, the traditional life of the man (outward-facing on the community and the world) and the traditional life of the woman (inward-facing on the home and family).

If we remove the latter, what choice have we but to substitute the former? There is nothing else.

In short, in the low-birth-rate world of the twenty-first century, one which is obviously necessary to survival, women must be allowed equal opportunity with men to enter into any branch of industry, politics, religion, science, or the arts that she wishes to. Women must have all the educational opportunities of men, all the economic opportunities, all the social opportunities. It must be a world in which women can make their own living and own their own souls.

And, if a woman does have a child (or, at most, two), she should not be penalized. She should not be forced to withdraw from the outer world and accept economic privation. In the low-birth-rate world, children, because they are few, will be extraordinarily valuable. They are the basic resource of the world and are of prime interest to society. If a mother wishes to devote herself fully to her child, that is a useful and important occupation and should be a well-paid one. If she has to work she must do (or wishes to do) in the world outside, there should be ample social arrangements for allowing her to do that, too.

Which is not to say that men must be left out of consideration. If women are free to adopt the outer-directed life, men are free to adopt the inner-directed life. In any association of two individuals, each can do the share of the work he or she would like to do; but the kind of work that neither wants to do must be equally shared, because the very concept of "men's work" and "women's work" will not exist, except where the unequal distribution of wombs and breasts forces a difference.

I don't view the kind of women's-equality world I have been describing as being wrenched from reluctant men by militant women; nor as being granted by selfless men to grateful women.

Quite otherwise. The women's-equality world will come about as the simple consequence of the type of society we will have in a low-birth-rate world. You can't have anything else. A low-birth-rate world *requires* women's equality. Without a women's-equality world, we can't have a low-birth-rate world.

And since it is quite clear that a low-birth-rate world is the price of the survival of our civilization, it follows that the acceptance of the ideals of women's equality is also the price of survival.

In fact, let's go a bit farther. Granted that A and B go together, which should come first? Ought we to wait until we have a low-birth-rate world before we strive to bring about a feminine-equality world, or should it be the other way around?

Suppose we first labor to bring about a low-birth-rate world? Is that possible under feminine-subservient conditions? Shall we attempt to persuade women not to have babies without showing them what the alternative is? Would they listen? And if they do listen, and the world becomes low-birth-rate, shall we then begin the slow job of selling feminine-equality, while women with nothing else to occupy them long to have children?

Or should it be the other way around? Should there be a strong effort to increase the value women place on themselves *right now,* even before the low-birth-rate world is achieved? Would this not encourage women to want to be more than baby machines? To see in the ideals of a low-birth-rate world a wider and more decent world for themselves? Or, if too late for themselves, at least one that will be better for their daughters? (It is taken for granted that men will endure much that the world might be better for their sons; why not women for their daughters?)

And as women are enlightened, will they not join in the struggle to reduce the birth rate? And if *they* want it reduced and become militant about it, will it not be difficult for men to say them nay?

It seems quite clear to me that if we try for a low birth-rate first and feminine equality second, we will get neither; while if we try for feminine equality first and get it, we will automatically get a low birth-rate as well.

And that is the clear and inevitable conclusion. Since we don't have much time, the price of survival is feminine equality as quickly as possible— even *now.*

*Afterword*: When a man writes an article like the above, it is inevitable that there may be a cynical murmur among the readers to the effect: "I wonder what *his* wife does?" Well, my wife is a physician who holds a responsible position at an important psychiatric institute. Her work is more specialized and difficult than mine is.

---

# 15

# Letter to a Newborn Child

Welcome, little child, to the four billion! That's how many there are of us on this planet—four billion.

We haven't always been four billion. Only fifty years ago or so, when I was born, there were only two billion people on earth. The population has doubled since then and it is increasing even faster now.

About one hundred twenty million people were born this year, and you were one of them. Of course about forty million people died this year, but that still means that the world's population went up by eighty million. And it will go up another eighty million next year, and eighty million more the year after. In fact, the additional number of people will itself go up from year to year.

By the time you are thirty-five years old there will be *eight* billion people in the world—unless something happens to prevent the increase.

One thing that could happen is that the rate at which people die will go up. That may very well happen, for mankind, in order to support his vastly increasing numbers, has been destroying his environment. He has been ransacking the earth for food, destroying the wilderness, and killing off the wildlife to make room for his herds and crops. He has been rifling the earth of its resources in order to obtain energy, metals, fertilizers, and material of all sorts to support his expanding number.

We have managed to keep pace with the population up to this point. Indeed, the four billion today are better off on the average than were the two billion who were here when I was born—but it isn't going to be possible to keep pace anymore. We have only been able to do it so far by continuing to destroy the very environment that supports us.

Now there is an energy crisis. It is difficult to get enough energy to keep everything going faster and faster and faster. It has become difficult to

72

keep the farm machinery in operation and difficult to make fertilizer cheap enough for poor nations to be able to afford it.

The weather is getting bad, too. Our industrial civilization is pouring more and more dust into the atmosphere, and this dust reflects more and more sunlight back into space. The earth has been cooling off, therefore, for the past thirty-five years. Not much—just a little—but it is enough to cut down the growing period and to alter the storm-tracks. We are having droughts, and crops aren't doing as well as they once did.

In addition, our industrial civilization produces pollution of all kinds that is helping to ruin the environment by poisoning the soil and the sea and thus destroying, little by little, the living creatures on which we live.

This means we will not be likely to increase the world's food supply in the coming years, and yet the world's population is still going up. Since we are already having trouble feeding the people of the world, we can be pretty sure that there are going to be famines in many parts of the world in the years ahead.

The nations that will be the first to suffer are those that are poor and already near starvation—and it is in just those nations that the population is going up most quickly. The chances are about 85 out of 100 that you were born in a poor nation—in Bangladesh, in India, in Indonesia, in Nigeria, in Paraguay, in Haiti. That means you are likely to be dead in a few years. Even if you live past childhood, you are likely to be hungry for all the years you spend on this planet.

And in the mad scramble for food on your part and on the part of billions of others, the people of earth will further damage the world they live in and will begin to fight each other over scraps. As things grow worse, the death rate will go very high, and all of civilization may crash. It means, then, that if you do manage to reach middle age you may find that the world is a savage place in which you and a few million others are living among the vast ruins of a richer time.

Is there anything that can be done to prevent this? Well, if we don't want to keep down the population by killing off people rapidly through famine and disease and war, the only alternative is to see to it that fewer people are born. The birth rate must go down.

You are a sweet and lovable child, as all children are, but there are too many of you. There *must* be fewer of you.

The birth rate may decline as people come to understand the deadly and present danger of population increase. In poor nations, the birth rate may decline if the standard of living can be made to rise. (For that reason the rich nations must do everything they can to help the poor ones—for their own selfish sake.)

If the birth rate goes down and if the population ceases to grow and even begins to decline, if we can have a world without war and irrational conflict, then human beings can perhaps turn their vast knowledge to the task of solving some of the enormous problems he has created.

Some of those problems have been created by the very technological expertise we have developed, and then unwisely used, but perhaps we have learned our lesson.

*Wisely* handled, we can use a still further advanced technology to prevent and reverse pollution, develop nonpolluting industries, preserve the beauty of nature and the cleanliness and purity of air, water, and soil. We can learn to conserve the limited resources of the world and distribute what we have more fairly so that as many people as possible can enjoy comfort and security. In a happier world, the threat of nuclear war may disappear, and new energy sources will be developed that will remove the threat of nuclear contamination.

It might be a better world eventually, after all—if not for you, then for your children's children. And with luck, you will be one of those who, when you grow up, will contribute to this better world.

We who went before you left the environment worse than we found it but perhaps you and those born along with you will leave it better than you found it, so that the human race and the world it lives in can be saved.

And if that is so, then welcome indeed.

# Part IV

# Science: Opinion

# 16

# Technophobia

There are a number of reasons why people might suffer from technophobia —
that is, a morbid fear of technological advance. Some have nothing to do
with technology itself, but merely with unpleasant consequences that might
equally well have arisen from other phenomena.

For instance, many people fear particular technological advances
because they see such things as possibly costing them their jobs. This is a
rational fear, and in a society that is indifferent to the plight of the jobless it
is even a justified one.

The same insecurity can, and does, bring about the fear of foreign com
petition, or of unrestricted immigration, or of a drive to employ hitherto ex-
cluded minorities. In every case, there is likely to be a push on the part of
those whose jobs are at risk for tariffs, quotas, or intolerance — not out of
economic or racial theories, but out of an understandable apprehension of
unemployment and hardship.

The treatment of this form of technophobia would require an enlightened social awareness that would allow, as one item, facilities for education
and retraining, since it is always certain types of jobs that shrink with
technological advance, never the total number.

A second reason for technophobia arises out of a lively appreciation of
the risks that accompany advances in technology: the risk of radiation as
nuclear power plants are built; the risk of poisonous wastes as factories of
various sorts proliferate; the risk of pollution and death that has arisen out
of our conversion into an automobile society. All this is reasonable enough,
although it usually requires a strong emphasis on the disadvantages, and no
consideration whatever of the concomitant advantages of technology. One
obvious way of fighting this form of technophobia is a constant *and visible*

preoccupation on the part of technologists with the control of such dangers, and with care for the environment.

Neither of these forms of technophobia is what I am dealing with now, however. Rather, I am concerned with a more subtle and less easily understood technophobia, one that involves the fear of a new advance in technology on the part of those whose professional career is actually involved with technology, and who clearly stand to benefit from the advance.

To be specific, we have entered the computer age, and there exist many devices that take advantage of computerization to allow a broad spectrum of machine capabilities that did not exist before. People can work with computers in such a way as to do much more with their telephones, their correspondence, their information storage and recall, and the control, regulation, and maintenance of their office procedures than would have been conceivable ten years ago.

Such computerization can only help the user, increase his responsibilities, and make him more valuable to his employer. There would seem to be no obvious disadvantage involved. Yet very many people resist new, technologically advanced equipment, and, if they are more or less forced to accept it, refrain from using it.

This is not something I know of only by hearsay. I am, myself, a prime sufferer from this form of technophobia, and in my case it is more puzzling (and more disgraceful) than in the case of others.

After all, it is not just that I would be benefited by accepting certain changes. I am actually an indefatigable propagandist for them. I have written numerous articles, for instance, in which I have praised the forthcoming computerization and automation of the world. I have considered it a desirable advance, and have argued in favor of accelerating it in every way. I have frequently pointed out the ways in which technological advance generally would serve to help solve the many serious problems that face the world today.

You would think then that I would consider it my duty to be in the forefront of change; to be forever ready to accept those advances that are involved in my profession and my way of life. Not so!

I am a writer, a prolific one. I have used a typewriter of one sort or another for well over forty years. All around me now, however, my fellow-writers are shifting to word-processors. I constantly heard of the advantages that word-processors have brought them, but I remained deaf to the lure. I clung to my typewriter.

Mind you, I could well afford to buy a word-processor, but I did not do so. I did not even investigate the situation. I made no effort to see what a word-processor looked like, how it worked, what it did. I preferred to pretend they did not exist and typed away stubbornly.

Came the day when a magazine editor asked me to write an article on my experiences with a word-processor. (Naturally, he assumed I had one and used it.) I had to admit I did not have one.

The editor took action at once. He obtained the cooperation of Radio Shack, and in a very brief time a TRS-80 Model II Micro-Computer, with a Scripsit™ program, arrived. The idea was that I was to have the word-processor set up; that I was to learn how to use it; and that I was then to write an article about it.

That seemed straightforward enough, but I was overcome with fear.

The Radio Shack people were enormously helpful, to be sure. Their representatives spent hours setting up the equipment and educating me in its use, but I reacted as though I were being invited into a dentist's chair. It was only shame that kept me from abandoning the project a dozen times over, and eventually I learned how to use the machine. I have used it steadily ever since, and, in fact, I am writing this essay on the word-processor.

But why, I ask myself, was I so resistant to something that was sure to help me (and does) and that did not even cost me anything to begin with?

My first argument was that I loved my typewriter and did not want to abandon it.

That was an outright lie. In the forty years during which I had used type-writers, I had switched from a manual to an electric without a pang, I had switched from an electric with a moving carriage to one with a moving "wal-nut" with delight. I had never suffered before when I abandoned an old, faithful machine. Why should I do so now? In fact, I haven't abandoned it. I still use my good old typewriter whenever I feel like it. I used it this morning to type some letters. Cross out sentiment, therefore.

A more rational argument I used was that, over the years, I had worked out a set of automatic motions that enabled me to type for endless hours without too much concern over details. Why should I disturb that situation?

There, perhaps, I had a point. I did not wish to give up something I had painfully learned, in order to start from scratch and learn something else. To be sure, I overestimated the difficulty involved. When I use the word-processor, I am still typing and making use of my automatic movements, for there is still a keyboard under my fingers. There are, however, new keys that I must use now and then and whose secrets I must learn. This proved not to be hard, and I can (and do) shift back and forth between my word-processor and my typewriter without a problem, but my fear of having to re-educate myself is worth looking into.

Human beings learn how to handle numerous complicated devices in their lifetimes. The learning is not always easy, but once the complications are learned—if they are learned properly—it all becomes automatic. The thought of abandoning it and learning something else, of going through the process *again,* is terribly frightening.

For instance, the system of common measures in the United States—inches, feet, yards, miles, or ounces and pounds, or pints, quarts and gallons—is an incredibly complicated and nonsensical farrago of units. The rest of the civilized world uses the metric system, which, in comparison, is simplicity itself. Using the metric system would save us endless hours of educating our youngsters and be beneficial to our entire industrial structure. There would be the initial expense of conversion, to be sure, but that would quickly be paid back by the savings that would ensue.

And yet there is no question that the American public fears the metric system and, if it had its will, it would cling to the present system forever. Nor is it because the public uses the common units with any great skill. Very few Americans are completely at ease with them, and know, offhand, how many pecks there are in a bushel, or how many square feet in an acre, or, for that matter, how many inches in a mile. Yet we won't change it for a system any child can learn in a day and remember for a lifetime. We invent reasons for resisting the change, but the real reason is that we dread the process of re-learning.

It is the same dread that keeps the world from adopting a sensible calendar in place of the one we now have, or of simplifying the spelling of English words. We insist on only twenty-eight days for February, and on spelling "nite," "night," simply because we *already know* the nonsense and we would have to *learn* the sense.

For that matter, millions of people have learned how to speed-type on the standard typewriter keyboard, which was worked up by the original inventor without much in the way of thought. It is possible to devise a keyboard with the letters so arranged that anyone's speed would be increased by at least 10 percent simply because the hands would have more nearly equal parts to play and there would be a greater proportion of shifts from one hand to the other and back. A new rational keyboard will not be accepted, however. There are all those millions who have invested in learning how to type speedily and automatically on the standard and senseless one.

There is your technophobia, then. It is the fear of re-education. Capable executives, educated and highly intelligent, are, if anything, more prone to it than others are; for in re-education you tend to start from scratch, and the high official actually has more baggage to throw out the window than people in lower ranks do. It is not pleasant to begin a process you had thought was over and done with forever, or to abandon a superiority that had been painfully achieved.

What does one do then? It would be ridiculous not to use new techniques that would clearly benefit us all. If we turn down the chance at a computerized society, as we have turned down the metric system, we will place our nation at an intolerable disadvantage with respect to other nations that have made the decision to forge ahead. We are already being shamed by Japan, which is making use of advances up to and including industrial robots. Then how do we overcome technophobia?

It is my feeling that re-education must be recognized for the highly difficult and (even more so) embarrassing process it is. In my case, my own techno-phobia with respect to the word-processor was overcome by the fact that the Radio Shack people came to my home and educated me in privacy. It was hard and embarrassing enough, to be sure, but I would never have made it if there were people around who knew how to use a word-processor and who watched me make a hash of it at first. Rather than experience the shame, I would have dropped the computer out of my thirty-third story window and cheerfully paid the damages.

I would suggest, then, that any person in an office who is required to use a device that is new to him be given an opportunity to learn how to do so *in private* and without undue hurry. There should never be any surprise at any difficulty in learning, nor any insistence that: "Really, the process is terribly simple." The process may indeed be simple, but the psychological difficulty of abandonment and re-starting is enormous and should be approached with sympathy.

If private and sympathetic re-education is the carrot, there is also the stick. Young people have no trouble in learning the new techniques. This is not because schoolchildren are brighter than adults, or because with age there comes a hardening of the mental arteries. Not at all! The youngster has not piled up a huge supply of knowledge and experience that he must unload and discard. Writing a theme on a blank page takes far less time than writing that same theme only after another theme already present (and written in more or less indelible ink) has been erased.

Nevertheless, the young people entering an industry can learn to use advanced techniques far more readily than old hands (who are superior in every way as far as experience and judgment is concerned) can do. One does not wish to throw out all that experience and judgment, but the executives who are being given a painstaking chance to learn the new techniques might as well know that, if all else fails, there *are* those ambitious young people waiting to climb the ladder.

For instance, I had no occasion to fear other writers learning to use a word-processor. A word-processor may make the mechanical side of writing simpler and faster, but it has no direct effect on quality; and as long as I felt that my stuff was *good,* no matter how old-fashioned the means of producing it might be, I could count on selling it and continuing to make a living. But suppose I got it into my head that one of the reasons editors like to deal with me is that I can turn out items quickly and that I can be counted on to meet my deadlines (which is true) and that I was going to be outpaced by other writers making use of word-processors and that editors were therefore going to become less interested in me. In that case, I would surely have turned to a word-processor much sooner than I did and would have faced the task of painful re-learning with a grimmer determination. Carrot and stick!

The task of re-education is, perhaps, not the only problem that stands in the way of executives wholeheartedly accepting the new technology.

In some ways, computerization eliminates the middleman. A properly programmed computer does many of the tasks that, by older ways, a secretary, typist, or file clerk would do. Throughout history, however, the status of a person has been measured (at least in part) by the number of other people who would jump to do his bidding. There is a tendency, therefore — whenever technological advance removes the need to use an underling — to use an underling anyway.

The telephone, for instance, makes it unnecessary to send a messenger. It is only necessary to lift a receiver and dial a number, and you can talk directly. There is nothing to the task, but the mere fact of doing so yourself cheapens you by the old standards. It therefore becomes necessary to ask your secretary to dial the number for you. If she encounters another secretary there may be a contest as to which party in the conversation loses status by getting on the telephone first.

Under these circumstances, it becomes difficult for an executive to make use of technologically advanced equipment without losing status. Striking a typewriter keyboard is a secretary's job even if the keyboard gives directions for an elaborate computerized system. Naturally, the more highly placed the executive, the greater the loss of status and the more reluctance there is to make use of the equipment. The tendency is for a new set of middlemen to be introduced — technicians who make use of the equipment on order. This, of course, increases expense and inefficiency.

How on earth can this be countered?

One way, perhaps, is to make use of another measure of status. Money!

In a computerized society, those who can make use of computers readily and easily are essential and should be paid accordingly. It seems reasonable to suppose that an executive who commands a substantially higher salary if he runs his own computers than he would if he used someone else to do so, would have a stronger tendency to use his own. Furthermore, if he valued the control of an underling more than the mere possession of money and was content with a lower salary, it might destroy some of the fun if that underling, handling the all-important computers, himself had an executive salary. It can become difficult to experience a true sense of superiority over someone else who displays an equal earning capacity.

This, however, could tend to be complicated. Trying to undo the abstraction of prestige by money alone might not always work. In any stratified society, there is always that peculiar attitude where a decayed "gentleman" who may be abysmally poor is nevertheless considered superior to a "tradesman" who is merely rich.

Fortunately, there is another route whereby the problem may be solved. Computers have made astonishing advances in the past forty years. Not even the most thoughtful science-fiction writer in 1942 would have dared describe computer systems of 1982 as they are actually proving to be.

And for a time, at least, it is reasonable to suppose that computers will continue to make rapid advances.

They will, it is likely, make advances in the direction of "friendliness"; that is, they will be improved in such a way as to make it easier for humans to operate them. The ultimate in ease of operation is through the use of speech as a mode of direction.

What distinguishes the human underling from all other devices whereby an executive can arrange to have an order followed is that the human being can obey a spoken command. In everything else, the executive must push a button, dial a number, write a note, abstract a card—do *something* other than speak.

If, then, the executive who wants a piece of information can say, "Get me thus-and-so," and the information is obtained, then it matters little whether the device that follows the order is a flesh-and-blood creature who places the information on the desk or a metal-and-electricity creature who displays the information on the screen. It will help, of course, if (supposing it to be the latter) the computer can say, "The information you requested, sir!" as the screen comes to life. If it can be programmed to say so in a respectful tone of voice, it will help even more.

Under such circumstances, someone using a computer system is not at all likely to feel much, if any, loss of status.

It will all come. Our increasingly complex civilization must be computerized if it is to continue to work, and technophobia cannot and will not be allowed to stand in the way. Therefore, we will have to understand the causes behind the reluctance to make the change and then take such action as will remove, or at least minimize, those causes.

# 17

# What Have You Done for Us Lately?

Technology is a favorite whipping-boy these days, and many who enjoy its benefits join in the outcry without for one moment letting go of those benefits.

We must understand what we mean by technology. As soon as human beings attempt to modify their environment to increase their comfort or pleasure by the use of anything other than their own unaided bodies, we have technology. The use of fire involved a series of technological advances. So did the use of clothing and of the simplest wood or stone tools. The lever, the wheel, the horseshoe, the magnetic compass, the printing press, and the clock are all examples of technology triumphant.

All right, say the technophobes (those who hate and fear technological advance), no one complains about those things. No one wants to go naked out into the world and scrabble to live with nothing but one's hands and feet. But what has technology done for us *lately*?

Shall we say the last two hundred years?

The coming of the Industrial Revolution lifted the burden of physical labor from the shoulders of humanity. In the pre-industrial world, 95 percent of the human race spent their short lives in unending digging, hauling, pushing, lifting—a way of life that was scarcely in any way different from that of the domestic animals they labored with.

The techniques of mass production made possible a supply of artifacts large enough to yield vast numbers of people serviceable material goods for their use, comfort, and amusement—houses, furnishings, dishes, tools, toys.

In less material ways, advancing technology made it possible to produce so much printed matter that mass literacy made sense, and for the first time vast numbers of people had an opportunity to participate in the intellectual sphere of life. In various ways, devised by technology, drama, music and art were brought to people who would never have dreamed of such a thing prior to the industrial age.

Medical advances, which would have been impossible without the help of technology, lifted from humanity the horrible dread of epidemic disease; removed the debility of vitamin shortages and hormonal imbalance; ended the torture of surgery without anesthesia; and, in short, doubled our life-span and increased a hundredfold our comfort and security.

Who wants to give all that up?

Ah, say the technophobes, but take a closer look at some of these so-called benefits. Look at the five-and-ten gimcrackery that machines turn out in place of the fine handcrafted ware of yore. Look at the trashy books, the cartoon art, the primitive rock-bands that replace the great literature, the fine design, the noble culture of the pre-industrial age.

It's an easy comparison to make if one chooses the worst of the present and contrasts it with the best (often an idealized and nonexistent best) of the past. The fact is that in pre-industrial times there was much junk manufactured, both materially and intellectually. It is the best that has, for the most part, survived; and that which has survived is idealized in the usual way critics have of praising the past for no other reason than that it is safely past.

Besides, the finely crafted ware, the great works of art, and all the rest of it were made for a thin scum of aristocracy that floated on the vast sea of miserable human beings for whom none of that existed. If we went back to careful hand-work, we'd return to art for the few and nothing for everyone else. After all, how many hands are skillful and how much can these hands do?

Do we now have gimcrackery and trash? Certainly—but we might improve it. And in any case it is better than the nothing of the pre-industrial age.

Even so, the technophobes might argue, how long can it all continue? What will technology do for us *now*?

We have advanced heedlessly and recklessly, they might say, at the cost of rifling the planet of its energy resources and polluting it very nearly to death. We have built our material power without advancing our wisdom, so that we may now destroy everything by nuclear war or by the collapse of a top-heavy centralized structure of this vaunted technology.

To an extent, that is true. But then even a rose garden has its thorns, so it should be no surprise that technological advance brings its problems.

Problems are meant to be solved, and while technology has no monopoly on the formation of problems, it has a near-monopoly on the solution of important problems. At least in the past it has almost always been technology that, in one way or another, has solved those problems. The solutions have, in their turn, created new problems, to be sure, but we must resolutely face that unavoidable fact and continue the search for further solutions.

What do we do about energy, for instance?

The technophobic solution might be to dismantle our nuclear power plants and switch to coal.

But coal-mining is dangerous and soil-destroying, coal-transportation is arduous, coal-burning fouls the air and kills more people than radiation ever did, and if coal is the major fuel it will, even more than oil, add steadily to the carbon dioxide in the air and alter the world's climate catastrophically.

The next technophobic solution is to abandon coal, too, and turn to the renewable energy resources: wood, wind, running water.

Alas, there isn't enough. The energy-consuming work of the world will wind down in that case.

The technophobic reaction might then be to let it wind down. Our forefathers got along without all that energy and were the better off for it, they might say.

But our forefathers in 1780, before the Industrial Revolution, numbered 900 million, and that number was maintained, for the most part, at subsistence level. Our own number in 1980 is 4 billion—five times as high. We can't very well go back to 1800 unless we're ready to reduce the world's population by 80 percent. How can we do that without catastrophe when no one is going to volunteer to go quietly?

What is the solution, then?

One is for new energy sources that can come only from a healthy technology moving forward. Nuclear fusion! Solar energy!

If we do want energy conservation (there's no harm in that), the proper route is by increasing the efficiency of use and distribution—through technology.

If we decide to place more weight on renewable resources, we must make use of new devices for the purpose or greatly increase the efficiency of old devices, or both—through technology.

If we decide that some energy resources are dangerous and yet cannot be dismantled, we must somehow make them safer—through technology.

If we decide to dismantle our centralized industrial apparatus because small is beautiful, we must do so in careful stages that will not, in the process, reduce the world to disaster and chaos. And how do we do that except by an advancing technology.

In fact, take it all in all, what are we arguing about? We cannot choose to abandon technology. That simply is not a rational choice. Even those apostles of medievalism, the Iranian ayatollahs who want to see all the world Islamic and all of Islam safely back in the seventh century, spread their word by television and defend their borders with jet planes and artillery.

All that the technophobes can do is get in the way and dishearten us and make it that much more difficult for us to solve our problems and march into the future.

They may ask, sardonically, of course, whether there is a future to march into, whether our problems may not have become insoluble.

Well, our problems may fail to be solved, thanks to a variety of human factors, among which are to be included the tactics of the technophobes. Nevertheless, they do not seem to be insoluble in principle, thanks to a still-advancing technology.

Computers are proliferating and growing more versatile. They can help us handle a world grown too large and complex to manage without them, to solve problems that have grown too subtle and deep to solve otherwise, to do work that could not be done as rapidly or as accurately under other conditions.

And our range has expanded. We have the technical capacity to move out into space.

There we can find energy from the sun in more copious quantities than on the earth. There we can find in the moon and the asteroids new and untouched reservoirs of material resources. There we can find properties of vacuum, zero gravity, high and low temperatures, hard radiation, around which we can build new laboratories and new factories. There we have the vast room into which to transfer much of the industrial world from the earth's surface, so that inevitable pollution can be discharged into space to be swept away by the solar wind; and so that work too dangerous for the earth can be carried out with the insulation of thousands of miles of vacuum protecting the population centers of the earth. There we can even have a new population outlet.

This will all bring us new problems, of course; but where problems exist, the response is solution or collapse. And if we choose solution, then it is technology or failure.

I choose solution, and I choose technology.

# 18

# Speculation

It is the business of a scientist to speculate, to hypothesize, to think of possible explanations—if not in public, then in the privacy of his own mind. In fact, a scientist cannot help doing so, any more than a writer can keep from thinking of fragments of plot or snatches of dialogue, or a musician can keep from hearing in his mind notes put together into themes and variations.

But when a nonscientist, a member of the general public, encounters scientific speculations, how can he tell whether there is something sound and possible in it or whether it is just nonsense?

It is quite likely that he *cannot* tell, simply because he lacks the necessary background, knowledge, and experience, any more than he can tell, when he lacks literary or musical talent of his own, whether some tentative literary or musical excursions bear promise of merit or not. He may know whether the writing or music sounds good and pleases him, but his personal taste is not the point at issue. I, for instance, am very fond of the works of Agatha Christie, and yet I strongly suspect they do not represent deathless prose.

In the same way, a scientific theory may please you and may seem in accord with your own feelings and beliefs, but your own pleasure is no proof of its possible validity.

What, then, does one do?

In the first place, one must consider the source. A recognized scientist is far more apt to produce a possibly fruitful piece of speculation than an unknown or an amateur is. Thus Francis Crick has speculated on life having originated on Earth through seeding (intentional or not) by extraterrestrial travelers; and Fred Hoyle has speculated that life can evolve in interstellar dust clouds or on comets and that the latter are the origin of earthly pandemics.

Neither speculation has (in my opinion) much chance of turning out to be useful; but, since Crick and Hoyle are first-class scientists, what they say cannot be dismissed out of hand. They are not likely to ignore obvious difficulties or to be unaware of the arguments that might be marshaled against their views. Taking such counterblows into skilled account, they keep their speculations from being easy to demolish.

Speculation by an unknown need not be taken so seriously as a rule, but the rule is not universal. Many a young man, unknown to fame, has come up with something that, in the end, made him world-renowned in science (sometimes after a delay of decades). Albert Einstein was a virtual unknown 26-year-old in 1905, when he advanced his theory of relativity, and many scientists refused to take him seriously at first.

Nevertheless, the chances that speculation by an unknown might turn out to be fruitful can be judged, at least to an extent, from its nature. There should be some indication that the speculator, however young and unknown, understands thoroughly the field he is dealing with—mathematics, physics, chemistry, medicine, whatever—and knows what has already been done.

The great scientific innovators of the past have always completely understood those aspects of science they overturned. Copernicus was a learned student of Ptolemaic astronomy, Galileo knew every aspect of Aristotelian physics, Vesalius had a thorough knowledge of Galenic medicine, and Einstein understood Newtonian physics completely. It is unlikely that anyone without thorough and evident comprehension of present-day scientific thought is likely to be truly innovative.

But who can judge adequately the background of the speculator? Again, qualified scientists are the only safe judges.

If a speculation bears promise of being useful because the speculator is properly grounded in the field, and, even better, if he has already demonstrated that grounding and has undoubted scientific standing, then we might term the speculation a hypothesis, but that in itself adds nothing. "Hypothesis," "speculation," and "thought" mean much the same thing, but the first is derived from Greek, the second from Latin, and the third from Anglo-Saxon, and the connotative prestige of each word is in accord with the "learned" nature of the language from which it is derived. A hypothesis is Greek enough to connote a thought that has the sanction of scientific experience on the part of the thinker.

If a hypothesis offers no way of being tested, with results that might either support or destroy it, then it can go no farther. It might be intellectually amusing or even stimulating, but it bears no promise of being useful.

If it can be tested, or better yet, if it can make predictions that no one would think of testing in the absence of the hypothesis, and if, as a result, one can better understand many observations, especially observations no one earlier would have thought to make, then the hypothesis becomes a "theory." (Note that a theory is *not* "just a supposition." It is a well-supported, well-tested, and well-accepted system of thought, which, if widely enough accepted, is sometimes termed a "natural law.")

By seeing the requirements of legitimate scientific speculation, one can get a notion of the reverse of the coin.

The signs of speculations and speculators that are not likely to be worth much are these:

They are advanced by people who have no standing in the field and who betray a lack of knowledge of work already done.

They make no use of standard terminology but make up their own terms, which are inadequately defined, and they do not use mathematical symbology where that would be expected. (Mathematics is almost impossible to fake, if you are not grounded in it.)

No adequate way of testing their suggestions is presented, and no useful predictions are made. Arguments are flawed, or unclear, to those educated in the field.

The speculator tends to be polemical and inordinately defensive. The speculator, lacking the knowledge that would enable him to build safeguards into his speculations that would guard against legitimate and foreseeable objections, and thus allow him to feel secure in his thinking, is apt to react angrily to unforseen or condescending objections (let alone abrupt dismissal) and is often so emotionally attached to his thinking that he suspects a persecuting conspiracy on the part of the "establishment." The speculator is so convinced of this that his continuing speculations often consist more of an attack on the "establishment" than a reasoned exposition of his own views.

None of this is perfect of course. Oliver Heaviside invented his own mathematical terminology, and Nikola Tesla was a polemicist with tendencies to paranoia, but both were great scientists. Still, by applying these criteria, you will detect the nonvalid speculators with very few exceptions.

An invalid or useless speculation, based on ignorance, is an example of "pseudoscience." "Pseudo" is from a Greek term meaning "false" or "deceiving." Pseudoscience is false science. It is nonsense that can confuse or mislead the unsophisticated, because it has some of the trappings of science, because it uses some of the language of science, because it deals with some of the interests of science, and because it calls itself science.

In fact, pseudoscientists, having no commitment to real science (or being unable to form such a commitment since they lack the knowledge and experience) tend to get satisfaction from the acclaim of nonscientists and to use that as compensation for the lack of appreciation they get from scientists. Either deliberately, or unconsciously, they tend to form their thoughts in such a way as to maximize that acclaim and, as a result, pseudoscience often becomes popular indeed with the nonscientific elements of the public.

Since these form the majority in terms of pure number, a pseudoscience like astrology is infinitely more popular than the true science of astronomy. Even the most quackish-sounding beliefs, such as pyramid power or the usefulness of talking to plants, quickly gain sway over the multitude. In fact, one might almost judge the worthlessness of a scientific speculation by the extent to which it gains a hold on the public.

Although pseudoscience is false science, if we reason etymologically, this is not to say that it is necessarily *deliberately* false science.

Many a speculator who produces what true scientists in the field would, almost unanimously, consider errant nonsense is nevertheless honestly and devotedly sincere in his beliefs. Such sincerity is not, in itself, evidence for the worth of the ideas, any more than popularity among the general public is. Nevertheless, sincerity is to be respected.

Remember, too, that pseudoscience may usefully stimulate scientific investigation and reasoning, even if only to develop arguments that will counter and demolish the nonsense in question. The effort, which might not otherwise have been made, would be a useful result of the pseudoscience, and one for which we ought to be grateful.

But what if one who promulgates a pseudoscientific speculation does so in the full knowledge that what he advances is nonsense? What if he does so merely to make money, or to gain power, or to play a practical joke, or to have the malicious fun of perpetrating mischief? For such purposes, he may even concoct or fake evidence and maintain it to be true.

What we have then are "hoaxes," and these, whether joking, malicious, or entrepreneurial, are always with us and must be guarded against.

And where, in all this, does my own specialty of "science fiction" fit in?

Science fiction is sometimes used as a synonym for pseudoscience, but this is quite wrong. Whereas pseudoscience passes for science, though falsely, science-fiction does not. Science fiction openly proclaims itself to be a product of imagination that bends its direction to the needs of science no more than it has to. And, because it is honest about this, science fiction is not a hoax, either.

Science-fiction writers speculate freely on scientific subjects, with the intention not of finding truth but of gaining a dramatic end. If, however, science-fiction writers are trained in science, they may find that their imagination is disciplined to the point where they cannot help coming across items of value.

Thus (to use myself as an example), my stories about robots, written in the 1940s, were unsophisticated in many ways, yet they contained enough to succeed in inspiring others who devoted themselves to the subject with great dedication and persistence, and who did indeed help to produce the industrial robots of today.

# 19

# Is It Wise for Us to Contact Advanced Civilizations?

It is always pleasant when people search for wisdom. How interesting to have a question start with "Is it wise—," rather than with "Is it profitable—" or 'Is it useful—" or "Is it safe—"

And yet profit, use, and safety all must go into a consideration of the wisdom of any course of action, so we might as well take these matters into consideration. To begin with:

1. *Is it profitable for us to contact advanced civilizations?*

At first thought, the only likely answer is, "No."

After all, to set up something like Project Cyclops, an array of more than a thousand 100-meter radio-telescopes, is going to take a great deal of cash. There is not only the material required for the building and the expense of construction, but also the money required for maintenance and for the salaries of the many people who will then be engaged in the project for, perhaps, years.

We are talking of billions of dollars.

The chances are high, moreover, that in return for those billions of dollars we will get absolutely nothing, in the sense that no advanced civilizations will be contacted.

This is true even though most astronomers are convinced that for various reasons such advanced civilizations exist, even perhaps in great numbers. After all—

a. Astronomers may be wrong. At some point or other in their chain of logic there could be a mistake and we may be the only civilization in existence.

b. Even if there are others, it may be that through the luck of the game there are none near enough to us in space to contact. We may be located in a galactic desert.

c. Even if there are relatively nearby civilizations, none may see any reason to spend a great deal of effort and energy on sending out signal beams at random through the universe. Certainly they would see no reason to spend that effort and energy to beam signals toward us specifically.

d. Even if the civilizations are advanced enough to make the sending out of signals a simple task for them, and even if they do it, it may well turn out that we lack the expertise to pick up the type of signal that they, with their advanced technology, choose to send out.

e. Even if we pick up their signal, it is not at all likely that we will be able to interpret it.

In short, in order to contact advanced civilizations, we must have one, or more, that is fairly close to us, who can send out signals and who choose to do so, who aims recognizable signals that move in our direction either accidentally or deliberately, and who sends signals that we can recognize and interpret.

Surely the chances of all that are so small that the expenditure of billions of dollars on the attempt is a wild and foolish extravagance.

—Except that science does not fit into such tight compartments. The choice is not: either make contact with advanced civilizations or nothing. We may fail to make contact and yet not end with nothing.

In the first place, the very attempt to construct the necessary equipment for Project Cyclops or anything similar will succeed in teaching us a great deal about radio telescopy and will undoubtedly advance the state of the art.

Second, it is impossible to search the heavens with new expertise, new delicacy, new persistence, and new power and fail to discover a great many new things about the universe that have nothing to do with advanced civilizations, whether we detect signals or not.

We can't say what those discoveries will be, in what direction they will enlighten us, or just how they may prove useful to us. However, knowledge, wisely used, has always been helpful to us in the past and will surely always be helpful to us in the future.

We must conclude, then, that the attempt to contact advanced civilizations is sure to be profitable.

2. *Is it useful for us to contact advanced civilizations?*

If the profitability of the search is going to arise serendipitously out of our blind scanning of the sky, without regard to whether we find the advanced civilizations or not, then why bother with those advanced civilizations? Will the attempt to find them just distract us from our real task of gathering knowledge?

After all even if the signals come in—

*a.* As I said before, they will be difficult or impossible to understand. How likely, after all, is it that we can penetrate the workings of an alien mind, when human beings have such trouble understanding one another?

*b.* Even if we manage to interpret the signals, they are bound to turn out to be trivial. They are, in all likelihood, sent out only to attract attention and to be an obvious indication of intelligent origin. We will end up, therefore, being told that $1 + 1 = 2$ and $2 + 2 = 4$. Interesting, but scarcely illuminating.

*c.* Even if, for any reason, the signals carry messages we can interpret and that *are* interesting to us, we can't really start a dialogue. The advanced civilization is bound to be a long distance away—fifty light-years is not an unreasonable guess at all; rather optimistic, if anything. That means that for any question sent out, an answer will be received a hundred years later.

Is there any answer to these very valid points?

Well—If we are capable of sending out a question, then the advanced civilization, on receiving it and knowing that there is intelligence on the other end, will perhaps at once begin transmitting in earnest. There may be a century wait to begin with, but thereafter we may get a cram course in all aspects of that alien civilization.

But what's the difference? Will we understand what we get?

Surely the question of understanding is not immediately vital. The task of deciphering the alien signals would be an interesting and challenging one and would be instructive in itself. We might gain some insight into alien psychologies in the process. Then even tiny breaks in the code would be of interest. Suppose all we can do is to pick up one hint that would help advance our knowledge of physics in some one respect. That one advance would not exist in a vacuum. We could pick it up and run with it on our own.

And, even if we never understand one thing about what they're saying—not one thing—the receipt of the signals is still important in itself, since it will tell us that an advanced civilization exists. There are very many reasons, after all, for wondering if civilization isn't a self-limiting thing, if any intelligent race is bound to learn to wield the forces of nature before they gather the wisdom to learn how to do so intelligently and who, therefore, proceed to destroy themselves.

We ourselves, alas, seem to be in the process of bringing about our self-destruction, and there are times when many of us must feel that the process is inevitable and can neither be stopped nor deflected, that a sure death-sentence hangs over us.

If we detect no civilizations, that doesn't mean they have all destroyed themselves, for there are many other reasons why we may not detect them.

If, however, we *do* detect an advanced civilization, then we know at once that at least one of the species has made it. And, if they, why not we? If we detect the least little signal of an alien intelligence—even if we cannot decipher one bit of it—we still have the knowledge that survival is possible and that, therefore, *we* may survive.

If all else fails, then the psychological value of contact is important and may prove even crucial to survival by helping to counteract despair.

It is clear, then, that to contact advanced civilizations cannot help but be enormously useful.

### 3. *Is it safe for us to contact advanced civilizations?*

After all, are we not safer in isolation? Is it wise to attract attention to ourselves? May not an advanced civilization, aware of our existence, send out their ships and take us over, exploit us, enslave us, wipe us out?

If we fear that, then we must also realize that we are no longer in isolation anyway and that it is too late to avoid attracting attention to ourselves. Ever since humanity has been using radio waves in quantity a sphere of radio-wave activity has been expanding out from Earth in all directions at the speed of light, and with steadily increasing intensity. An advanced civilization may pick it up and, even if they can't interpret the details of the signals, they'll know we are here.

To be sure, the involuntary sending out of messages is excessively weak and at the distance of even the nearest civilization may be too weak to be detectable. Ought we to make things worse by sending out deliberate signals?

At the moment, that is not in question. All we are engaging to do is to receive signals, to *listen*. We are free not to answer any signals if we choose not to.

But what if we *do* decide to answer? Is *that* safe?

Consider that, if there are advanced civilizations in the universe, then some of them may be very old. The universe and our own galaxy have lasted long enough to contain civilizations that are as much as ten billion years old. In ten billion years, civilizations will certainly have explored our entire galaxy, will have recorded every planet capable of supporting life, and, if that represented their choice of action, would have colonized them all.

The mere fact that humanity is here on this planet and that, as nearly as we can tell, life has developed, undisturbed, for over three billion years indicates that such a conquering galactic civilization does not exist.

Why not? It may be that (a) civilizations, however old, cannot leave their home planets, (b) they can leave but do not choose to, (c) they have left, but believe in allowing life-bearing planets to develop their own intelligent life-forms free of interference.

The best reason for civilizations', however old, not being able to leave their home planets rests with the speed-of-light limit to travel. If there is no way of getting round the speed-of-light limit, then it would take hundreds or thousands of years to travel from one habitable world to the next, and this is not an attractive prospect. Each civilization would then limit its expansion to the neighborhood of its own planet. In fact, the mere existence of signals would indicate that the civilization sending them feels nailed to the spot and can reach out only by speed-of-light radiation.

Even if the speed-of-light limit is not absolute and if there are ways of getting round it, the difficulties may be too great to allow for the kind of mass transfer of populations that would be involved in conquest and settlement. It may be that civilizations would use it only as a means for sending out scouting vessels to explore and to gain knowledge of the universe. Such scouting vessels might have noted Earth's existence thousands of years ago, before civilization appeared on Earth. We would be viewed not as a world for settlement but as a world for interesting observation; and, if we find that signals seem to be aimed at us particularly, that may be the reason.

Finally, even if advanced civilizations find methods for making flights between the stars as simple as we find flights between cities, then this does not necessarily mean they would conquer us.

We know from our own experience how extraordinarily contentious and quarrelsome the members of an intelligent species can be. We also know how difficult it is to make major advances, such as those required in the exploration of space, when the various segments of our species spend almost all their time, money, and effort in quarreling with each other. In fact, it doesn't really seem likely that humanity will be able to advance into space unless the peoples of Earth abandon war and agree to make the advance a truly cooperative venture. Space exploration is a global concern and can only succeed if it is a global activity.

We might argue, therefore, that any intelligent species that cannot control its contentiousness will destroy itself before it goes out into space (as we may). On the other hand, any intelligent species that makes its way out into space succeeds in doing so only because it isn't contentious in the first place, or has learned to control its contentiousness if it is. It will therefore be more likely to seek a League of Galactic Civilizations than to attempt conquest.

For all these reasons—because the advanced civilizations can't get at us; because, if they can, they are surely peaceful; because, if they can get at us and are not peaceful, we've given ourselves away anyhow—we conclude that it is safe (or, at any rate, involves no additional risk) to contact advanced civilizations.

Finally, since it is profitable, useful, and safe to contact advanced civilizations, there is no possible conclusion that we can come to but that it is wise to contact advanced civilization.

In fact, it would be very unwise not to.

# 20

# Pure and Impure

It is easy to divide a human being into mind and body and (if you are an intellectual) to attach far greater importance and reverence to the mind. To be sure, philosophers must eat, but that might be viewed as a regrettable necessity to be neutralized by edifying conversations conducted across the dinner table.

Similarly, the products of the human mind can be divided into two classes; those that serve to elevate the mind and those that serve to comfort the body. The former are the "liberal arts"; the latter, the "mechanical arts."

The liberal arts are those suitable for free men (from the Latin *liber* meaning "free") who are in a position to profit from the labors of others in such a way that they are not compelled to work themselves. The liberal arts deal with "pure knowledge" and are highly thought of, as all things pure must be.

The mechanical arts, which serve agriculture, commerce, and industry, are necessary, too; but, as long as slaves, serfs, peasants, and others of low degree know such things, educated gentlemen of leisure can do without it.

Among the liberal arts are some aspects of science. Surely a study of the complex influences that govern the motions of the heavenly bodies and control the properties of mathematical figures and of the universe are pure enough.

As time went on, though, science developed a low habit of becoming applicable to the work of the world and, as a result, those whose field of mental endeavor lies in the liberal arts (minus science) tend to look down upon scientists as being in altogether too great a danger of dirtying their hands.

Scientists, in response, tend to ape this Greek-inherited snobbishness. They divide science into two sorts. One deals only with the difficult, the abstruse, the elegant, the fundamental — in other words, "pure science," a truly liberal art.

Clearly, any branch of science that goes slumming and becomes associated with such disregarded mechanical arts as medicine, agriculture, and industry is a form of impure science.

"Impure" is a rather pejorative adjective. It is more common to talk of "basic science" and "applied science."

On the other hand, differentiation by adjective may not seem enough. The same noun applied to both makes the higher suspect and lends the lower too much credit. There has therefore been an increasing tendency to call applied science "technology."

We can therefore speak of "science" and "technology" and we know very well which is the loftier, nobler, more aristocratic and (in a whisper) the purer of the two.

Yet the division is man-made and arbitrary and has no meaning in reality. The advance of knowledge of the physical universe in all its aspects rests on science *and* technology, and neither can flourish without the other.

Technology is indeed the older of the two. Long before any human being could possibly have become interested in vague speculations about the universe, the hominid precursors of modern human beings were chipping rocks in order to get a sharp edge, and, with that, technology was born.

Further advances, by hit and miss, trial and error, and even by hard thought, were slow, of course, in the absence of some understanding of basic principles that would guide the technologists in the direction of the possible and inspire them with a grasp of the potential.

Science, as distinct from technology, can be traced back as far as the ancient Greeks, who advanced beautiful and intricate speculations. The speculations tended to become perhaps more beautiful with time and certainly more intricate, but there was no way in which they could have become more in accord with reality. The Greeks, alas, spun their speculations out of deductions from what they guessed to be first principles, and they sharply limited any temptation to indulge in a comparison of their conclusions with the world about them.

It was only when scientists began to observe the real world and to manipulate it that "experimental science" arose. This was in the sixteenth century, and the most able early practitioner was the Italian scientist Galileo Galilei (1564-1642), who began work toward the end of that century. Thus began the "Scientific Revolution."

In the eighteenth century, when enough scientists recognized their responsibility toward the mechanical arts, we had the "Industrial Revolution," and that reshaped human life.

Such is the psychological set of our minds toward a separation of science into pure and impure, basic and applied, useless and useful, intellectual and industrial that even today it is difficult for people to understand the strong union of the two branches of science and to grasp the frequent and necessary interplay between them.

Consider the first great technologist of the modern era, the Scottish engineer James Watt (1736-1819). Though he did not invent the steam engine, he developed the first one with a condensing chamber and was the first to devise attachments that converted the back-and-forth motion of a piston into the turning of a wheel. He also devised the first automatic feedback devices that controlled the engine's output of steam. In short, beginning in 1769 he developed the first truly practical and versatile device for turning inanimate heat into work, and with that began the Industrial Revolution.

But was Watt a mere tinkerer? Was he a technologist and nothing more?

There lived at the time a Scottish chemist, Joseph Black (1728-1799), who, in his scientific studies of heat in 1764, measured the quantity of heat it took to boil water. As heat poured into water, its temperature went up rapidly. As water began to boil, however, vast quantities of heat were absorbed without further rise in temperature. The heat went entirely into the conversion of liquid into vapor and this is "the latent heat of evaporation." The result was that steam contained far more energy than did hot water at the same temperature.

Watt, who knew Black, learned of this latent heat and familiarized himself with the principle involved. That principle guided him in his improvements in the then-existing steam-engines. Black, in turn, impressed with the exciting application of his discovery, lent Watt a large sum of money to support him in his work.

The Industrial Revolution, then, was the product of a fusion of science and technology.

Nor is the flow of knowledge entirely in the direction from science toward technology. While many people (even nonscientists) can now recognize that scientific research and discovery, however pure and useless it may seem, may turn out to have some impure and practical application, few (even among scientists) seem to recognize that, if anything, the flow is stronger in the other direction. Science would stop dead without technology.

Galileo, in 1581, when he was seventeen years old, discovered the principle of the pendulum. In the 1590s, he went on to study the behavior of falling bodies and was greatly hampered by his lack of any device to measure small intervals of time accurately. No such device then existed.

The first good timepiece was developed in 1656, by the Dutch scientist Christian Huygens (1629-1695), who applied Galileo's principle of the pendulum to construct what we would today call a "grandfather's clock."

The principle of the pendulum, by itself, would have done little to advance science. The application of the pendulum principle and the technological development of timepieces made it possible for scientists to make the kind of observations they could never have made before so that science, even the purest, could leap ahead.

In similar fashion, astronomy could not possibly have progressed much past Copernicus without the intervention of technology. In fact, without technology Copernicus' principle of sun-at-center could never have been firmly established in the place of the ancient Greek Earth-at-center principle.

The crucial key to astronomical advance began with spectacle-makers, mere artisans who ground lenses, and with an idle boy-apprentice who, in 1608, played with those lenses—and discovered the principle of the telescope. Galileo built such a telescope and turned it on the heavens, and no greater revolution in knowledge has ever happened in so short a time as the second it took him to turn his telescope on the moon and see mountains there.

In fact, the history of modern science is the history of the development, through technology, of the instruments that are its tools.

Nor is that the only influence of technology. The products of technology offer a field for renewed speculation.

For instance, although Watt had greatly increased the efficiency of the steam engine, it still remained very inefficient. Up to 95 percent of the heat energy of the burning fuel was wasted and was not converted into useful work.

A French physicist, Nicolas Carnot (1796-1832), applied himself to this problem. Involving himself with something as technological as the steam engine, he began to consider the flow of heat from a hot body to a cold body and ended up founding the science of thermodynamics (from the Greek for "heat-movement") in 1824. In fact, he described a version of what we now call "the second law of thermodynamics," one of the great triumphs of pure science—which grew out of the impure science of the steam engine.

Nor was it only in the past that science and technology interacted. The interaction has grown constantly stronger, and the two have never been so intertwined as now.

The year 1979 was, by coincidence, a significant year for two great men who seem to typify the very epitome of the purest of science on the one hand and the most practical of technology on the other—Albert Einstein (1879-1955), the greatest scientist since Newton, and Thomas Alva Edison (1847-1931), the greatest inventor since anybody. How did the work of each invade the field of the other?

Surely the theory of relativity, which Einstein originated, is as pure an example of science as one can imagine. The word "practical" seems a blasphemy when applied to it.

Yet the theory of relativity describes, as nothing else can, the behavior of objects moving at sizable fractions of the speed of light. Subatomic particles move at such speeds, and they cannot be studied properly without a consideration of their "relativistic motions."

This means that modern particle accelerators can't exist without taking into account Einstein's theory, and all our present uses of the products of these accelerators would go by the board. We would not have radioisotopes, for instance, for use in medicine, in industry, in chemical analysis—and of course we would not have them as tools in advancing research into pure science either.

Out of the theory of relativity, moreover, came deductions that interrelated matter and energy in a definite way (the famous $e = mc^2$). Until then, matter and energy had been thought to be independent and unconnected entities.

Guided by that equation, the energy aspects of research in subatomic particles were made more meaningful and, in the end, the nuclear bomb was invented and nuclear power stations were made possible.

Einstein worked outside the field of relativity, too. In 1917 he pointed out that, if a molecule were at a high-energy level (a concept made possible by the purely scientific quantum theory, which had its origin in 1900) and if it were struck by a photon (a unit of radiation energy) of just the proper frequency, the molecule would drop to lower energy. It would do this because it would give up some of its energy in the form of a photon of the precise frequency and moving in the precise direction as the original photon.

Thirty-six years later, in 1953, Charles Hard Townes (b. 1915) made use of Einstein's theoretical reasoning to invent the "maser," which could, in this way, amplify a short-wave radio ("microwave") beam of photons into a much stronger beam. In 1960, Theodore Harold Maiman (b. 1927) extended the principle to the still shorter wave photons of visible light and devised the first "laser."

The laser, based on Einstein's abstruse reasoning of four decades earlier, has infinite applications, from eye surgery to possible use as a war weapon. It can bounce back into the realm of pure science, too, for it can be used for unprecedentedly delicate experiments that serve to verify the theory of relativity.

And Edison?

The net result of his inventions was to spread the use of electricity the world over, to increase greatly the facilities for the generation and transmission of electricity, to make more important any device that would make that generation and transmission more efficient and economical. In short, Edison made the pure-science study of the flow and behavior of the electric current an important field of study.

Charles Proteus Steinmetz (1865-1923) was certainly a technologist. He worked for General Electric and he had two hundred patents in his name.

Yet he also worked out, in complete mathematical detail, the intricacies of alternating-current circuitry, a towering achievement in pure science. Similar work was done by Oliver Heaviside (1850-1925).

As for Edison, his work on the electric light unwittingly led him in the direction of purity.

After he had developed the electric light, he labored for years to improve its efficiency and, in particular, to make the glowing filament last longer before breaking. As was usual for him, he tried everything he could think of. One of his hit-and-miss efforts was to seal a metal wire into the evacuated light bulb near the filament, but not touching it. The two were separated by a small gap of vacuum.

Edison then turned on the electric current to see if the presence of the metal wire would somehow preserve the life of the glowing filament. It didn't, and Edison abandoned the approach. However, he noticed that an electric current flowed from the filament to the wire across that vacuum gap.

Nothing in Edison's vast practical knowledge of electricity explained that, but he observed it, wrote it up in his notebook and, in 1884, patented it. The phenomenon was called the "Edison effect," and it was Edison's only discovery in pure science—but it arose directly from his technology.

Did this seemingly casual observation lead to anything? Well, it indicated that an electric current had a flow of matter of a particularly subtle sort associated with it—matter that was eventually shown to be electrons, the first subatomic particles to be recognized.

Once this was discovered, methods were found to modify and amplify the electron flow in vacuum and, in this way, to control the behavior of an electric current with far greater delicacy than the flipping of switches or the closing of contacts could. Out of the Edison effect came the huge field of electronics.

There are other examples. A technological search for methods to eliminate static in radio-telephony served as the basis for the development of radio astronomy and the discovery of such pure-science phenomena as quasars, pulsars, and the big bang.

The technological development of the transistor brought on an improved way of manipulating and controlling electric currents and has led to the computerization and automation of society. Computers have become essential tools in both technology and science. A computer was even necessary for the solution of one of the most famous problems in pure mathematics— the four-color problem.

The technological development of a liquid-fuel rocket has led to something as purely astronomical as the detailed mapping of Mars and of experiments with its soil.

The fact is that science and technology are one!

Just as there is only one species of human being on Earth, with all divisions into races, cultures, and nations, but man-made ways of obscuring that fundamental truth; so there is only one scientific endeavor on Earth—the pursuit of knowledge and understanding—and all divisions into disciplines and levels of purity are but man-made ways of obscuring *that* fundamental truth.

# 21

# Do We Regulate Science?

There's a strong impulse to attempt to control the direction of scientific research because in recent years there has been a steadily intensifying notion that science can and perhaps will present humanity with overwhelming danger.

Who can feel at ease as scientists learn more and more about nerve gases, sophisticated space weapons, and genetic engineering? Who can be satisfied with the pile-up of radioactive wastes, the prospects of ever more deadly war, the possibilities of modifying human structure and behavior?

Why should we not establish a review board to consider the routes along which scientific research is progressing — to slow them here, hasten them there, turn them this way or that in another place, and, on occasion, to stop a particular line of research cold?

But it's not so easy. The curious human mind makes strange leaps and a discovery aborted here is then duplicated there. In 1847, an Italian chemist, Ascanio Sobrero, discovered nitroglycerine and (inevitably) also discovered its explosive quality. Horrified at the destructive uses to which he foresaw it might be put, he stopped all research in that direction.

It didn't help. Others made the same discovery, and those others did not stop.

For that matter, should it have been stopped? Certainly advances in the knowledge of explosives produced new and deadlier weapons by the end of the century. On the other hand, Alfred Nobel tamed nitroglycerine and produced dynamite, and there's no need to go into all the constructive uses of high explosives.

We must, in other words, make a distinction between knowledge itself and the uses to which knowledge is put.

Almost any piece of knowledge can be used in what would seem to be both constructive and destructive ways, and this is nothing new. In prehistoric times, stone axes and stone-tipped spears made it possible for human beings to face the larger predators with greater chances of survival, and they also made it easier for human beings to maim and murder other human beings. The ability to start a fire at will yielded people the advantages of cooked food, pottery, glass, and metals — and the disadvantage of accidental conflagration or deliberate arson.

Even the development of speech brought with it, aside from its obvious advantages, a new and more sophisticated level of possibility for lying and deceit.

There is no question, then, but that humanity must at all times question and inspect the uses to which knowledge is put, and this, in actual fact, is done. From earliest times, a major effort of government has been to regulate the activities of human beings in such a way as to minimize harm. If this has not always worked perfectly, it is because of deficient information and because of the human passions of greed and hate, which have always excused deliberate harm to the persons and properties of those seen as enemies — or even as just strangers.

Nowadays, because our ability to do harm has grown greater, because we have learned that even clearly constructive uses may have unexpectedly harmful side-effects, because our technological society has made us all so interdependent that the very concepts of "enemy" and "stranger" have lost their meaning, we are condemned to strive harder to try to foresee and avert danger.

None of this, however, can or should imply that the acquisition of knowledge itself must be regulated, directed, or stopped. Knowledge increases options, offering us additional opportunities to manipulate the universe for good or for evil, and, if we choose wisely, we end with more opportunity for good.

Thus, to know the nature and uses of vitamins is to give us the opportunity of improving nutrition generally, while also offering us the risk of hypervitaminoses. We can avoid hypervitaminoses by increasing our knowledge of vitamins further and learning the dangers of overdoses; or we can avoid hypervitaminoses by not having learned about vitamins in the first place and risking the possibility of making such avitaminoses as scurvy, rickets, and pellagra into scourges. History shows us which is the wiser move.

We might argue that in the case of vitamins, the dice are loaded in favor of knowledge. In the case of nuclear energy, however, the dice seem to be loaded the other way. To be sure, our increasing skill in making use of nuclear reactions have given us radioisotopes for biochemical research and radioimmunoassay for diagnostic procedures, but it has also given us the nuclear bomb and radioactive wastes. Would we not lose the former cheerfully, if it meant we would be freed of our fear of the latter? Would not ignorance have been bliss in that case?

Would it? Even where new knowledge offers little good and much evil, might we not select the little good and discard the much evil? Or is humanity so certain to choose the evil out of some kind of malevolent stupidity that ignorance is the only way out?

If the latter is true, nothing will save us. A humanity intent on destroying itself if it has the opportunity to do so will be just as intent on finding the opportunity in the first place, and there is no use in further argument. We are doomed.

If, however, we do have the faculty of intelligent choice, then let us make that choice as effective as possibly by constantly increasing knowledge of the potential dangers to be avoided as well as of the usefulness to be chosen.

Is it better to shield a child from all harm by imprisoning it in a nursery with padded walls or to bring it out into the world, teaching him or her to recognize danger and to learn how to avoid it? If you have had a child, which course did you follow?

Or consider this— Of all scientific advances, those in medical science most nearly meet with universal approval. Who would object to a cure for cancer, for instance?

The most important single advance in medical science was the development of the germ theory of disease by Louis Pasteur in the 1860s. Thanks to this, infectious disease was controlled and the plagues and epidemics that threatened humanity all through history dwindled and began to vanish. Life expectance doubled from 35 years to 70, and who can complain about that?

Except that with the decline in the death rate, the population explosion was fueled and world population has quadrupled since Pasteur's time. Right now, the world's steadily increasing population threatens us all with destruction. It aggravates every ill we experience—declining resources, increasing pollution, alienation, crime, terrorism, and war.

Ought we then to have stopped Pasteur? Ought we to have forbidden the germ theory as dangerous knowledge designed to destroy civilization in a century and a half?

Or ought we to have told him to go ahead, to improve individual health and life, and to have worked steadily toward population control through a lowering of the birth rate?

# 22

# For Public Understanding of Science

Leon E. Trachtman, in a recent essay, questioned the assumptions underlying the usual consensus that keeping the public informed on science is a
good thing in a democratic society. He lists the assumptions as follows:

1. Knowledge is simply a good thing in itself.
2. People will be able to make more intelligent, personal consumer decisions if they have more knowledge of science and technology.
3. The very structure of a democratic society depends upon the existence
of an enlightened citizenry. The political and social behavior of this citzenry in
voting, in influencing elective and appointed officials, and in engaging in
political and social activism, will be more constructive for society if it is informed by solid scientific understanding.

As far as assumption one is concerned, Trachtman says: "With this
claim I have no argument, but it can hardly be the basis for the expenditure
of hundreds of thousands of dollars annually as part of a deliberate policy
of informing the public about science."

In a nation whose chief executive has just advanced a budget that
allocates a quarter of a trillion dollars in one year to weapons of war, a few
hundred thousand dollars for science education would seem trivial,
especially if the other two assumptions hold up, but we all have our
priorities, and I hesitate to expect others to live with mine.

As far as assumption two is concerned, Trachtman feels that the vast
amount of information spewed forth by the media, much of it self-contra-
dictory, leaves the public confused and unable to come to any decisions as
to what to buy or how to live. He says, concerning the citizen, "If he were
completely uninformed and simply followed the advice of his doctor

or an appropriate government agency to eat a moderate and balanced diet, he would almost certainly be as well off."

Perhaps! And yet for the first time in history, smoking is in decline. The full force of the law could not bring about a decline in drinking in Prohibition days — rather the reverse — but the repeated reports on the undeniable connection between smoking and lung cancer and heart disease are actually producing some results. Combine that with the publicity on cholesterol and it seems that the cardiovascular death-rate has been declining for a number of years now.

Thousands of lives have surely been saved, but whether that is worth the expenditure of hundreds of thousands of dollars depends on one's priorities. *I* think so.

As for assumption three, Trachtman feels that a confused citizenry is not likely to become activist, and that if they do become activist they will, with annoying stupidity, champion the wrong causes. Furthermore, popular treatment of scientific material will oversimplify, select for sensationalism's sake, report prematurely, fail to get across an understanding of the scientific method, raise false expectations, and do other vicious things.

This implies that the only kind of scientific information that the public will ever get will proceed from those who are basically as ignorant as the public is. What Trachtman complains of is the result, not of science education, but of the *failure* of science education. It is (or should be) the aim of those interested in communicating science to educate not only the public, but (even more important) those people in the media who report on science. It is for precisely this reason that a new organization, Science in the Public Interest (SIPI), has come into existence.

And are Trachtman's three assumptions the only ones that can be offered as a rationale for the scientific education of the public? Of course not! Confining it to those three implies an additional assumption that science education is intended for the good of the public alone. As it happens, science education is essential to the welfare of science and scientists *as well,* and with that in mind I offer three more assumptions of my own:

4. Science and scientists need the sympathy of the public.

Suppose scientists take the attitude: We are the scientific elite and you, the ignorant laymen, might just as well remain ignorant because you will not listen, and, if you listen, you will not understand.

In that case, scientists will become a priesthood, and the public (not quite as ignorant, or as stupid, as the elitists may think) will both hate and fear them — and perhaps with reason. Scientists who will not trouble to state their case, explain their findings, and show a decent and visible interest in the public and its good, will find themselves denounced and persecuted. The damage done to nuclear technology over the years is a case in point.

5. Science and scientists need the financial support of the public.

It would be pleasant if all scientists could support their work out of their salaries, but they cannot. Virtually all science needs the support of an academic institution, a private corporation, or the government. Academic institutions rarely have enough money; private corporations are naturally interested in solutions to specific problems; and only the government can support many aspects of science that are vitally important, but of no visible profit-generating potential in the near future.

Government money comes out of the wallets of the taxpayer and, if the public is not informed about science, does not see its value, and is not impressed with the motives of scientists, those wallets will button tightly. Already Senator Proxmire makes endless political capital out of deriding and persecuting science that he does not understand, and there will be millions of Proxmires if scientists retire haughtily to their ivory towers.

6. Science and scientists need to recruit their numbers from the public.

Scientists do not reproduce by binary fission; nor do they invariably give birth to children who are cut out to be scientists. My father was certainly no scientist; nor was any ancestor of mine as far as I know. I was recruited because I was fascinated by the science books I read as a youngster, and I have received countless letters from people who tell me that they were recruited through having read one or another of *my* science books for the public. This, in itself, would be sufficient if it were all that science education performed. Let scientists withdraw to their proud self-containment and let recruitment dry up, and science will wither quickly enough.

One thing is true. Attempting to educate the public in science *is* difficult. It is hard enough to get the essence of science across to graduate students, let alone to people who have never learned the art of rational thought.

The stakes, however, are very high, and we have no choice but to try—and, as we try, to endeavor to learn how to try even harder and better—and to remain undaunted by defeat.

We may, in the end, lose. We may, in the end, have accomplished nothing and left the world uninformed after all. We may (as Trachtman gloomily suspects) merely succeed in confusing the public, at the cost of hundreds of thousands of dollars (half an advanced warplane) a year.

But what is the alternative? To abandon the fight? To hold high the tattered banner of defeat? To leave the world to the *National Enquirer,* the astrologers, and the creationists? Shall we march into the darkness loudly crying, "We give up. They are just as well off ignorant, anyway. And at least we save a lot of money and in two years we can buy one more beautiful warplane."

Never! As for myself, I may be defeated at last, but I intend to struggle to the end. I will not surrender, embrace ignorance, and kiss its hideous face.

# 23

# Science Corps

Japan, with half the population of the United States, graduates five times as many engineers each year. For every American high-school student who has taken calculus, there are fifty Soviet students who have.

The world is moving into a high-technology future even while the United States, almost alone among the developed nations, is moving backward into scientific illiteracy. A shrinking scientific elite will be trying to keep the United States in the technological race as other nations pass us by and leave us behind.

What are we to do?

Scientific decay is not something we can reverse overnight. We can solve some parts of the problem (such as shortages of supplies and equipment in our schools) by throwing money at it, but the nation is in no mood to throw money at anything but "defense," and it does not see a scientifically literate population as part of that.

We can change this attitude, explain the need to approach crucial decisions with widespread understanding of the scientific issues involved in energy, food supplies, pollution, ecological balance, and so on.

This will take time, but no matter how much time it takes the world of American science must strive toward it. We must learn how to talk to the public, how to make our case, how to stress our importance to the very life of the nation. We must interest private industry in the importance of education if the government is blind to it, we must seek for closer cooperation between industrial laboratories and academic ones, we must ourselves go into the grade schools and public schools. And while we are doing all this — anything else?

Perhaps we can come up with a new concept. Twenty years ago, we established the Peace Corps, a group of willing amateurs who would labor for a better life in the undeveloped nations. Might we not now come up with a Science Corps, a group of willing amateurs who would have minds and hands that could help in scientific work?

At least one science, astronomy, has a long-established tradition of amateurism. There are amateur astronomers who sweep the stars in the search for comets and asteroids, whose photographs can be of the greatest value, who work in the tedious lower reaches of the science so that the professionals can remain at the cutting edge of the unknown.

Is there no possibility that this can be done in physics or chemistry or biology or geology as well? There are bright youngsters in high school and even primary school, self-educated beyond their grades, who would welcome a chance to work in any of a hundred fields, under the direction of experienced scientists, in return for the additional education they would receive in the process and the honor they would get for what they would accomplish.

There are men of middle age, earning their living in any of a myriad of nonscientific classifications, who are self-educated in one or another branch of science. That self-education might be incomplete, but might they not be willing to put it to what use is possible for the thrill of being involved in the real world of scientific investigation?

The amateurs of the Science Corps could not only do work of value, while freeing the more formally educated members of the scientific community to do the more difficult jobs and to be the decision-makers, but the Science Corps could itself become a great educational force. Its members would not be part of a suspect "priesthood"; they would be ordinary members of society whose very existence would bring science closer to the level of the people.

Their own enthusiasm, fresher and less jaded than that of the professionals, would be catching. They could go into the schools with less condescension and more excitement than a professor could, and be followed more eagerly. Youngsters who would doubt their own ability (or desire) ever to become a stuffy professor might find it easy to identify with one of their own.

And this could trigger off a beneficial cycle. The more popular (and populist) science becomes, the more willing the general population will be to support it. The more labor-intensive science can become through the hands of willing amateurs, and the less capital-intensive, the more science will be perceived as a bargain, returning more than it charges, and again, the more willing the general population will be to support it.

The more interested the population becomes in science, the more willing the government will be to use public money for the purpose, since no backlash will be feared. And as money flows in, science will expand and grow healthier. Education will improve and students will want and will get science courses. They can swell the rank of the Science Corps further, and it may become a worldwide phenomenon that will help the entire planet march into the supertechnological future of computers and space.

---

24

---

# Science and Beauty

One of Walt Whitman's best-known poems is this one:

*When I heard the learn'd astronomer,*
*When the proofs, the figures, were ranged in columns before me,*
*When I was shown the charts and diagrams, to add, divide and measure them,*
*When I sitting heard the astronomer where he lectured with much applause*
  *in the lecture-room,*
*How soon unaccountable I became tired and sick,*
*Till rising and gliding out I wander'd off by myself,*
*In the mystical moist night-air, and from time to time,*
*Look'd up in perfect silence at the stars.*

I imagine that many people reading those lines tell themselves, exultantly, "How true! Science just sucks all the beauty out of everything, reducing it all to numbers and tables and measurements! Why bother learning all that junk when I can just go out and look at the stars?"

That is a very convenient point of view since it makes it not only unnecessary, but downright aesthetically wrong, to try to follow all that hard stuff in science. Instead, you can just take a look at the night sky, get a quick beauty fix, and go off to a nightclub.

The trouble is that Whitman is talking through his hat, but the poor soul didn't know any better.

I don't deny that the night sky is beautiful, and I have in my time spread out on a hillside for hours looking at the stars and being awed by their beauty (and receiving bug-bites whose marks took weeks to go away).

113

But what I see—those quiet, twinkling points of light—*is not all the beauty there is*. Should I stare lovingly at a single leaf and willingly remain ignorant of the forest? Should I be satisfied to watch the sun glinting off a single pebble and scorn any knowledge of a beach?

Those bright spots in the sky that we call planets are worlds. There are worlds with thick atmospheres of carbon dioxide and sulfuric acid; worlds of red-hot liquid with hurricanes that could gulp down the whole earth; dead worlds with quiet pock-marks of craters; worlds with volcanoes puffing plumes of dust into airlessness; worlds with pink and desolate deserts—each with a weird and unearthly beauty that boils down to a mere speck of light if we just gaze at the night sky.

Those other bright spots, which are stars rather than planets, are actually suns. Some of them are of incomparable grandeur, each glowing with the light of a thousand suns like ours; some of them are merely red-hot coals doling out their energy stingily. Some of them are compact bodies as massive as our sun, but with all that mass squeezed into a ball smaller than the earth. Some are more compact still, with the mass of the sun squeezed down into the volume of a small asteroid. And some are more compact still, with their mass shrinking down to a volume of zero, the site of which is marked by an intense gravitational field that swallows up everything and gives back nothing; with matter spiraling into that bottomless hole and giving out a wild death-scream of X-rays.

There are stars that pulsate endlessly in a great cosmic breathing; and others that, having consumed their fuel, expand and redden until they swallow up their planets, if they have any (and someday, billions of years from now, our sun will expand and the earth will crisp and sere and vaporize into a gas of iron and rock with no sign of the life it once bore). And some stars explode in a vast cataclysm whose ferocious blast of cosmic rays, hurrying outward at nearly the speed of light reaching across thousands of light years to touch the earth and supply some of the driving force of evolution through mutations.

Those paltry few stars we see as we look up in perfect silence (some 2,500 no more on even the darkest and clearest night) are joined by a vast horde we don't see, up to as many as three hundred billion—300,000,000,000 —to form an enormous pinwheel in space. This pinwheel, the Milky Way galaxy, stretches so widely that it takes light, moving at 186,282 miles each *second,* a hundred thousand *years* to cross it from end to end; and it rotates about its center in a vast and stately turn that takes two hundred million years to complete—and the sun and the earth and we ourselves all make that turn.

Beyond our Milky Way galaxy are others, a score or so of them bound to our own in a cluster of galaxies, most of them small, with no more than a few billion stars in each; but with one at least, the Andromeda galaxy, twice as large as our own.

Beyond our own cluster, other galaxies and other clusters exist; some clusters made up of thousands of galaxies. They stretch outward and outward as far as our best telescopes can see, with no visible sign of an end—perhaps a hundred billion of them in all.

And in more and more of those galaxies we are becoming aware of violence at the centers—of great explosions and outpourings of radiation, marking the death of perhaps millions of stars. Even at the center of our own galaxy there is incredible violence masked from our own solar system far in the outskirts by enormous clouds of dust and gas that lie between us and the heaving center.

Some galactic centers are so bright that they can be seen from distances of billions of light-years, distances from which the galaxies themselves cannot be seen and only the bright starlike centers of ravening energy show up—as quasars. Some of these have been detected from more than ten billion light-years away.

All these galaxies are hurrying outward from each other in a vast universal expansion that began fifteen billion years ago, when all the matter in the universe was in a tiny sphere that exploded in the hugest conceivable shatter to form the galaxies.

The universe may expand forever or the day may come when the expansion slows and turns back into a contraction to re-form the tiny sphere and begin the game all over again so that the whole universe is exhaling and inhaling in breaths that are perhaps a trillion years long.

And all of this vision—far beyond the scale of human imaginings—was made possible by the works of hundreds of "learn'd" astronomers. All of it; *all* of it was discovered after the death of Whitman in 1892, and most of it in the past twenty-five years, so that the poor poet never knew what a stultified and limited beauty he observed when he "look'd up in perfect silence at the stars."

Nor can we know or imagine now the limitless beauty yet to be revealed in the future—by science.

# 25

# Art and Science

Knowledge is indivisible. When people grow wise in one direction, they are sure to make it easier for themselves to grow wise in other directions as well. On the other hand, when they split up knowledge, concentrate on their own field, and scorn and ignore other fields, they grow less wise—even in their own field.

How often people speak of art and science as though they were two entirely different things, with no interconnection. An artist is emotional, they think, and uses only his intuition; he sees all at once and has no need of reason. A scientist is cold, they think, and uses only his reason; he argues carefully step by step, and needs no imagination.

That is all wrong. The true artist is quite rational as well as imaginative and knows what he is doing; if he does not, his art suffers. The true scientist is quite imaginative as well as rational, and sometimes leaps to solutions where reason can follow only slowly; if he does not, his science suffers.

If we go through the history of human advance, we find that there are many places where art and science intermingled and where an advance in one was impossible without an advance in the other.

In early modern times, for instance, artists tried to work out ways in which to make the scenes they drew look more like the world they were trying to imitate. They drew on a flat surface, but they wanted to make their scenes look as though they had depth and "perspective."

To do that, they had to make some things look smaller in a very careful way. An Italian artist named Leone Battista Alberti published a book in 1434 in which he showed artists how to work out perspective properly. To do so, however, they had to use mathematics. It turned out that Alberti, in

116

working on a purely artistic problem, had developed the beginnings of a very important branch of mathematics called "projective geometry."

Again, in the Middle Ages, the knowledge of human anatomy was small because it was forbidden to dissect dead human bodies. Since a knowledge of anatomy was important if medicine were to advance, medicine did *not* advance for centuries.

But a proper knowledge of anatomy is also important in art. An Italian artist named Leonardo da Vinci wanted to draw human figures that looked real, and for that purpose he had to know how the bones and muscles inside the body were organized. About 1500, he dissected some thirty dead bodies, studied their muscles and bones, and drew beautiful pictures of them. He also studied the structure of the heart and from that got a notion of how the blood circulated.

It worked the other way around, too. A half-century later, a Belgian physician, Andreas Vesalius, dissected human bodies and, in 1543, published a great book on the subject, called *On the Structure of the Human Body*.

This was the foundation of modern anatomy and, in some ways, the foundation of modern medicine. Yet Vesalius was not the only one in the field. Other physicians were also dissecting, and they too were publishing books on anatomy. What was it that made Vesalius the greatest of these?

Art!

Vesalius commissioned a Dutch painter, Jan Stevenszoon van Kalkar (a disciple of the great Venetian painter Titian), to illustrate the book. No number of words can describe an anatomical structure as well as a beautiful picture can, and it was the illustrations more than the words that made Vesalius the "father of anatomy."

The connection between art and science continued in later times, too. In 1801, a German scientist, Johann Wilhelm Ritter, found that sunlight broke up a white compound called silver chloride and formed tiny black grains of metallic silver.

Since sunlight thus turns white to black, can sunlight be used to paint a picture? Scientists did not tackle this problem, but an artist did. He was a Frenchman named Louis Jacques Mandé Daguerre, who painted scenic backdrops for theatrical performances. He wondered if he could make those backdrops more realistic if he used sunlight to produce a light-dark pattern mechanically, a pattern that was exactly like that of something real. In the 1830s, he began to produce the first primitive photographs.

How could science do without photography these days? Astronomy would stop dead in its tracks if it couldn't photograph the heavens. Where would medicine be without X-ray photographs?

For that matter, photography has become a beautiful art-form in its own right, and all kinds of scientific advances have succeeded in making it more so. Chemicals that react more rapidly to light make short-exposure photographs possible. Special dyes make color photography possible. New mechanical devices make motion pictures possible.

Over and over again, modern scientists make great leaps into new realms of knowledge by looking upon the universe with the eyes of artists. They can't help but assume that the universe works symmetrically, that its machinery is orderly and beautiful and simple. They have faith that an explanation that has artistic beauty is more likely to describe the universe accurately than one that has not. A solution of artistic beauty is called an "elegant" one, and all scientists search for elegance.

The Scottish scientist, James Clerk Maxwell, for instance, by 1879 worked out four equations that could be expressed simply and neatly and which worked together with great symmetrical beauty. They were elegant. These equations described all the phenomena that had been observed in connection with electricity, magnetism, and light. This persuaded scientists that the equations were true and useful, but their elegance helped make them acceptable, too.

Since then, other great scientific theories have caught the imagination of the world because important concepts could be expressed in a few simple symbols. An important concept of the quantum theory is expressed as $e = h\mathrm{n}$, and an important concept of the theory of special relativity can be expressed as $e = mc^2$.

The theory of general relativity, first worked out by Albert Einstein in 1916, is still not completely acceptable. There are alternative theories advanced by other scientists. It is very difficult to make the necessary observations that will enable scientists to choose among them. Of all the theories, however, Einstein's is the simplest and neatest; it is the most elegant. Many physicists are sure it is the correct one because it is the most artistic.

In 1874, a Dutch chemist, Jacobus Henricus van't Hoff, worked out a theory that finally explained many of the problems that had puzzled chemists about the complex molecules of living tissue. Each carbon atom could attach itself by four "bonds" to four other atoms, and van't Hoff worked out the "tetrahedral carbon atom." The four bonds, he showed, were in the directions of the vertices of an imaginary tetrahedron surrounding the carbon atom.

It was a very elegant way of explaining many problems. What's more, molecules could be drawn three-dimensionally and they became art-forms as well as scientific facts. Eventually, in 1953, James Watson and Francis Crick worked out the double-helix structure of the nucleic acids, the key molecules of life, working from certain symmetries that had been observed about them.

Each year, the *McGraw-Hill Yearbook of Science and Technology* publishes a selection of the photographic highlights of the year, photographs made for scientific purposes that nevertheless have beauty and artistic value as well.

If you look at an electron micrograph of a sponge spicule or of a diatom (you can find both in the 1977 *Yearbook*), you don't know whether to admire them as products of science or as works of artistic beauty.

—And it doesn't matter; the two are the same.

---

## —— 26 ——

# The Fascination of Science

Carl Sagan's television series "Cosmos" offers us something unusual—a view of science in a grand sweep from the most ancient speculations we know of to the most modern discoveries we have made, and making use of the most advanced television techniques to lure us into understanding.

It offers us something even more unusual than that—it offers us the sight of an audience of millions of people who will eagerly watch a view of science that is not watered down.

As a spinoff from the television series there will be Sagan's *Cosmos,* a book that will reduce to print the words, vision, and action of the series. The first printing is 150,000 and there is no doubt that a number of printings will be required.

To be sure, Sagan is an attractive, well-spoken, brilliant person, a professional astronomer of imagination, capacity, and renown, and a writer of great skill—so one might presume that it isn't science the public is watching and reading, but Sagan.

That would be a tempting conclusion were it not for the fact that we are seeing an explosion of science magazines on the newsstands—magazines of real science, for the most part, that resist the temptation to fade off into mysticism and fairy tales.

We are seeing also a steady rise in the popularity of science fiction. This is true in the printed medium, where, in previous decades, science fiction had been the least-regarded of the categories of popular fiction, but where now it grows steadily while other branches of fiction are withering. It is spectacularly true in the visual media, where the great Hollywood blockbusters of the contemporary scene are "space operas" and where "The Empire Strikes Back" had even blasé observers cheering from their seats.

In fact, if I may (more or less blushingly) refer to personal evidence, among my 218 books published over the past thirty years, there are some 50 books of science fiction and 120 books of science fact and, far from striking any point of diminishing returns, I find that (so far) they are steadily doing better as the audience grows.

Why this fascination on the part of the public with science? And why now?

To be sure, there have always been people fascinated by science, to the point, in some cases, of finding nothing else of real value in life. These have always been few in absolute numbers and vanishingly small as a percentage of the population. Now, however, the numbers are increasing remarkably, explosively, and science is becoming almost a mass preoccupation.

Again, why?

I use the word "fascination" deliberately. It is derived from a Latin word meaning "spell." Something is fascinating that seems to absorb you beyond what would seem natural, that holds you enrapt and seems to deprive you of the ability or the will or the desire to break away. We use the word, usually, in a pleasant sense. One is fascinated by great beauty, grace, intelligence, picturesqueness.

But the pleasant sense is not obligatory. Among our nature myths is that of the mouse fascinated by the glittering eye of the snake, cowering helplessly, and waiting to be eaten. We can be fascinated by evil, by danger—unable to break away until it is too late.

It is fascination in this double sense that connects science and the general population.

It was not always so. Europeans and Americans who lived through the industrializing era of the nineteenth century were dimly aware of the existence of science in much the way they were dimly aware of the existence of China. What really affected their everyday lives and roused their liveliest interest was "invention."

They were perfectly aware of the changes produced in society and in their daily lives by such items as the steamship, the locomotive, the telegraph, the telephone, the electric light, the sewing machine. These were not, in the general view, the products of science but of the ingenuity of clever men who were not pictured as scientists (and who, indeed, were *not* scientists in the narrow sense of the word).

Yet there were scientists like André Marie Ampère, who worked out the mathematics of electrodynamics a generation before Thomas Alva Edison used electrodynamics in many of his inventions. For every person who has heard of Ampère, one can easily imagine that 100,000 have heard of Edison. What's more, of those who have heard of Edison very few understand the connection with Ampère, or see that Ampère had to precede Edison, for it was Ampère who made Edison possible.

The general understanding that science, and not necessity only, is the mother of invention is undoubtedly a twentieth century phenomenon, and so is the realization that science can be an instrument of destruction and retrogression as well as of advance and progression.

It may well be that the first glimmer of the meaning and potentialities of science (as opposed to "invention"), as far as the general public was concerned, came in 1915, with the use of poison gas in World War I. This was clearly a scientific development, pure chemistry; and it was a horrifying discovery without socially redeeming value, since it didn't even win the war for either side. Before 1915 was over, both sides were using it and it gave neither side an advantage but merely introduced a vast increase in the terror and misery for soldiers on both sides.

The terror wasn't forgotten. Poison gas wasn't used in World War II because it would gain nothng for either side but retaliation, yet civilian preparations always included the inevitable gas-mask.

And, even if it had been forgotten, World War II brought the nuclear bomb in 1945. Before it, even poison gas shriveled as horror, and the nuclear bomb, even more clearly than the earlier terror, was the product of science.

In the generation since World War II, science has continued to produce its marvels (and its horrors). Television and jet planes are inventions in the full nineteenth-century sense, but it is now clear that electronics and aeronautics are sciences and that the public knows the connection.

The development of solid-state physics has brought the transistor and all its infra-miniaturized descendents, and given us generation upon generation of computers in rapid succession, each set smaller, cheaper, and more versatile than the one before. And it would take someone quite incredibly naive to think of these computers as the product of mere ingenious tinkerers.

We cannot dismiss space exploration with the thought that rockets are no more than an invention of the medieval Chinese. A rocket, no matter how large and powerful, would merely be a missile that went up and never came down (at least not for years). What counts is the telemetry, the miniaturized devices, the solar cells that make the satellite or the probe responsive to our orders and humble transmitters of information.

Nor is there any question in the mind of much of the public that if our problems are to be solved — or made worse — it will be through the medium of science and technology. It doesn't matter whether a person is pro- or anti-technology, the realization remains if the person is not a total dreamer.

If the energy crisis is to be solved by the discovery and utilization of new sources (fusion? solar power stations in space? geothermal energy? biomass?), it is the advance of science that will make that possible and practical.

If the energy crisis is to be solved by the abandonment of "big science" and the development, somehow, of "people's science," of small backpacks of something or other, of one-man solar devices, of backyard steel-forging, of careful recycling of human wastes, we nevertheless have 4.2 billion people in the world who can't be allowed to starve by the hundreds of millions if we expect civilization to survive. To switch from the large-is-efficient to the small-is-beautiful will *still* take careful scientific and technological advance.

The American public is even aware that one of the components of American power in the world has been its leadership in science and technology — computers, microelectronics, subatomic physics, lasers, and so on — not only as a matter of war-weapons but as the backbone and basis of an advanced and productive industry. They are aware that the decline in American power is, in part at least, brought about by our diminishing status as world scientific leader.

In short, since 1945 the public view of science has changed. Science is no longer a remote discipline practiced by absent-minded professors and oddballs with long hair who speak a language none can understand but themselves, and whose conclusions, even when made partially comprehensible, clearly possess no importance whatever compared to tomorrow's football game.

Science is increasingly viewed, instead, as a matter of life and death to each one of us, and scientists are saviors/destroyers whom it is important to understand, and who must be brought into the marketplace in order that they might acount for what they are doing and that they might be told what to do next.

The French politician Georges Clemenceau, in one of his best-known bits of phrase-making, said, "War is too important to be left to the generals." This can be broadened to read: "Any specialty, if important, is too important to be left to the specialists."

After all, the specialist cannot function unless he concentrates more or less entirely on his specialty and, in doing so, he will ignore the vast universe lying outside and miss important elements that ought to help guide his judgment. He therefore needs the help of the nonspecialist, who, while relying on the specialist for key information, can yet supply the necessary judgment based on everything else — provided the nonspecialist can understand the specialist in the first place.

Science, therefore, has become too important to be left to the scientists. Scientists must be guided by a smoothly functioning society that rests upon an informed public opinion.

Every one of us has a life-and-death stake in science, every one of us has the responsibility and duty of helping to make decisions as to what problems science should tackle; what precautions science should take; how, and in what way, and where new scientific discoveries should be applied or not applied. And none of this can be directed out of ignorance or prejudice, but only out of understanding and wisdom.

Can such general public understanding and wisdom be gained and made use of? Clearly this is not something easily achieved, but just as clearly the first step is to learn as much about science and its current state as possible, and I suspect that more and more people are beginning to think so.

It is perhaps for this reason, therefore, that more and more people are interested in watching and reading science — real science — explained in terms that the nonspecialist can understand.

How does all this apply to science fiction? Science fiction, after all, is not science. At best, science fiction contains a leavening of science that can form only a minor portion of the whole, since the interest in any story is bound to be concentrated on the people in it; on their deeds and their reactions.

Such science as is included or discussed in science-fiction stories may, moreover, be oversimplified, modified, or distorted for the sake of the plot. For that matter, the science may (alas for human frailty) be downright wrong because of the ignorance of the writer — who is rarely himself a scientist.

It may be, then, that science fiction is gaining in popularity for reasons that have nothing to do with the popularity gain of science-for-the-layman.

At the start, that must certainly have been true.

Consider that important social change is always brought about by advances in science and technology. Other kinds of changes — the deaths of kings, the falls of dynasty, the sweeps of conquest or of pestilence — bulk large in the immediate event, but once the change has settled and the tide has receded, human beings go on to live as before. For that reason, the writer of the biblical book of Ecclesiastes felt justified in moaning, "There is no new thing under the Sun."

But compare such trivial and temporary changes with the permanence of the effects on every aspect of life of such things as the taming of fire, the development of agriculture, the invention of writing, the coming into use of pottery and metals, the discovery of the magnetic compass or of printing or — in more recent times — the coming into being of the steam engine, the automobile, television, the jet plane, the computer.

Scientific and technological advance is cumulative and accelerating. Each advance makes further advance easier and serves as a basis for still larger advances.

At the start, the rate of change through science and technology was so slow that the amount of significant social alteration in the course of a single lifetime was small enough to be unnoticeable, so that the wail of Ecclesiastes must have seemed true to individuals.

As the centuries passed, however, the rate of advance increased and the drumfire of change hastened its beat. Finally, about 1800, the rate of change, in those parts of the world where science and technology were advancing most rapidly, became great enough to be visible in the course of a single lifetime.

Human beings could observe what difference had been introduced in their own lifetime by the coming of the steam engine, for instance, or the development of gas lighting.

That created a new curiosity; possibly the only basically new curiosity to be introduced in historic times —

"What will life be like after I die?"

Prior to the nineteenth century, no one would have dreamed of asking that, since life would differ in the future in only inconsequential details from the past, as far as people could see.

By the nineteenth century, however, the question had meaning. What new inventions would appear? What new scientific discoveries? What new basic changes in life-style?

Science fiction arose in response to such questions. If one could not witness the future and assuage one's curiosity directly, one could at least speculate. Those who could speculate best, most eloquently, and most convincingly, did it professionally for those who could not.

The first true science-fiction writer — that is, the first to make a good living out of the craft — was Jules Verne. In the more than a century that has elapsed since his first success, those who followed him have been emulating him in speculating on future developments and on change.

As time has continued to move on, as the nineteenth century has faded into the twentieth, as the twentieth approaches its final decades, the rate of scientific and technological advance has continued to increase. Each change has followed closer and closer on the heels of the other, until it seems all but impossible to absorb them all. More and more it has become the fundamental crisis of our time that we may lack the ability to understand and accept change.

Unfortunately, change is always difficult to accept. We grow accustomed to whatever evanescent and unimportant ways and customs surround us as we mature, and thereafter that is our standard of "normal," "good," "eternal," and all deviations therefrom (the most necessary as well as the most pernicious) are resisted.

Yet, while change may be disliked and resisted, it *will* come; and if, as a last resort, it is stubbornly ignored, it *will* overwhelm us. Like it or not, change must be a factor in our calculations, and young people particularly are becoming increasingly aware of that.

It may be that that is why there is a curious flavor of decay and irrelevance in mainstream "realistic" fiction of today. As long as fiction deals with the here-and-now, the young people of today have to recognize it as nothing more than quaint. Perhaps that is why most forms of popular fiction have been decaying for a generation; why there is little fiction in the magazines; why the short story of almost all varieties is all but dead; why first novels are harder to publish than ever before.

It cannot be the effect of television alone, for in this same period, science fiction (science fiction *in print*) has been steadily expanding — in short-story form as well as in novels.

It is not that science fiction is an accurate way of predicting the future. The predictive record of the science-fiction writer, while better than that of almost anyone else, is still poor. Nevertheless, one thing *every* science-fiction story takes for granted is that the future will be different from today and that particular prediction, at least, is a remarkably certain one.

It is that basic assumption that makes science fiction distinctive that also makes it relevant.

It isn't at all likely that science-fiction readers in general have reasoned this out and have therefore become science-fiction readers. It is much more likely that hardly any of them have. Nevertheless there must be a general unease in the air today that bears the stamp "inevitable and continuing change" upon it. People must feel that this is the mark of the age, even if they don't think about it or put it into words, and must be drawn to that form of literature that bears the same mark.

It can be concluded, then, that the increasing tendency to be interested in science fact and science fiction is indeed part of the same phenomenon — the desire to accept and understand and, therefore, just possibly to *guide* change, both with the mind (science fact) and the heart (science fiction.)

But with all this really help us guide change? Will it teach us to solve the formidable crises of our times?

Perhaps not, but as the sad old joke has it: It couldn't hurt!

# Sherlock Holmes as Chemist

We all know that Sherlock Holmes was the first important detective in fiction to go about his business with true scientific rigor. At least we all think we know that. Arthur Conan Doyle wrote the sixty novels and short stories about the master with such winning conviction that he succeeded in convincing his readers that this was so.

Yet that conviction is an illusion. Conan Doyle was surprisingly poor in science, apparently, and Sherlock Holmes, as a scientific detective, does not really come off well for that reason.

Conan Doyle's limitations are visible in his attempt to describe the scientific profundities of the arch-villain James Moriarty, for instance.

In "The Final Problem," Holmes says of Moriarty, "At the age of twenty-one he wrote a treatise upon the Binomial Theorem, which has had a European vogue."

Moriarty was 21 years old in 1865 (it is estimated), but forty years earlier than that the Norwegian mathematician Niels Henrik Abel had fully worked out the last detail of the mathematical subject known as "the binomial theorem," leaving Moriarty nothing to do on the matter. It was completely solved and has not advanced beyond Abel to this day.

Then, in "The Valley of Fear," Holmes says of Moriarty, "Is he not the celebrated author of 'The Dynamics of an Asteroid'—a book which ascends to such rarefied heights of pure mathematics that there was no man in the scientific press capable of criticizing it?"

Why the dynamics of *an* asteroid, when there were already hundreds known in Moriarty's day? In the Newtonian sense, there was nothing further to be done about asteroidal motion after 1825, when the French astronomer Pierre Simon de Laplace completed his book *Celestial Mechanics*. To be sure, Moriarty might have anticipated Einstein's theory of relativity, or he might have solved what is called the "three-body problem" in gravitation, something that remains unsolved to this day. In either case, however, the work would have had general applications and would have applied to *all* moving bodies, and not merely to "an Asteroid."

But let us ignore mathematics and astronomy, which, we may fairly assume, was not Conan Doyle's forte. Let us, instead, turn to chemistry. Conan Doyle was a physician and one cannot have been a physician, even a hundred years ago, without *some* acquaintance with the principles of chemistry.

And it is chemistry that is the true test; for, if Conan Doyle portrays Sherlock Holmes as anything other than a detective of superlative genius, it is as a chemist. That makes sense, too, for chemistry has great forensic value and would be of prime importance to a scientific detective.

In "A Study in Scarlet," the first tale of the series, in which the meeting and first acquaintanceship of Holmes and his ever-after-loyal-companion-and-Boswell, Dr. John H. Watson, is described, we learn about Holmes's intellectual attainments. Watson makes a list of them and does so without pity.

He describes Holmes's knowledge of literature as "nil" and uses the same word for his knowledge of both philosophy and astronomy. Holmes's knowledge of politics is "feeble," his knowledge of botany "variable," his knowledge of anatomy "unsystematic," his knowledge of geology "limited."

When it comes to chemistry, however, Dr. Watson characterizes Holmes's knowledge of the subject to be "profound." We are therefore entitled to believe that Holmes is an expert chemist, and that Conan Doyle should labor to make him appear to be one.

And yet, although Conan Doyle dutifully mentions Holmes's chemical labors in a number of stories, he also manages to be wrong in one respect or another in virtually every case.

For instance, in "The Adventure of Shoscombe Old Place," Holmes says, speaking of the police, "Since I ran down that coiner by the zinc and copper filings in the seam of his cuff they have begun to realize the importance of the microscope."

It would seem that Holmes made a microscopic study of the dust gathered from the seam and detected metallic particles that he identified as zinc and copper. It would be an easy task to spot the metallic particles, but to identify them as being of this specific metal or that by eye alone is much trickier. No chemist would be satisfied with only visual evidence in such a case; certainly the courts would not. As it happens, even small quantities of copper and zinc could be tested for chemically, and the spectroscope was already in use and that would make the matter certain. Yet Holmes does not mention such tests.

If there is some possibility that chemical or spectroscopic tests were done but not mentioned, there is the fact that a few paragraphs earlier Holmes makes another kind of identification by microscope alone. He says of the material he is examining, "Those hairs are threads from a tweed coat. The irregular grey masses are dust. There are epithelial scales on the left. Those brown blobs in the centre are undoubtedly glue." And it is the glue that is the essential clue.

This is miraculous. To look at tiny blobs of amorphous organic material and to be able to tell that they are glue rather than any of a large variety of other amorphous organic materials represents particularly piercing eyesight. Holmes advances this identification as proving the guilt of a man suspected of committing a murder. If the courts accepted such evidence, there would none of us be safe.

But then, Holmes's eyes are such that, as he explains in "A Study of Scarlet," "I can distinguish at a glance the ash of any known brand either of cigar or of tobacco." If he can, he is the only human being on Earth or in history who can or could.

The first time Watson sets eyes on Holmes in "A Study of Scarlet," Holmes is working in a chemical laboratory and has just made an important discovery. Holmes cries out, "I have found a reagent which is precipitated by haemoglobin, and by nothing else."

The test is never referred to again in either this story or any of the 59 that followed, but a certain reasonable latitude for the imagination is permissible. What happens afterward is considerably less permissible, however.

Holmes offers to demonstrate the new test by pricking his finger with a "bodkin" to obtain some blood. He draws off "the resulting drop of blood" and adds it to a "litre of water." He then successfully performs the test that demonstrates the presence of the small quantity of blood in that large quantity of water. (To make the value of the test obvious, he ought to demonstrate that the reagent does *not* react with other substances that resemble blood in appearance, but we'll ignore that point.)

A drop of water is usually taken to represent a volume of about 1/20 of a milliliter. Blood, being more viscous, is likely to form a larger drop, but let us suppose that Holmes squeezes out just a tiny bit of blood and not a full drop and that he adds but 1/50 of a milliliter to the water.

A milliliter is 1/1000 of a liter, so 1/50 of a milliliter is 1/50,000 of a liter. In adding the blood to the water, a proportion of 1 part of blood to 50,000 of water is produced, yet Holmes says, "The proportion of blood cannot be more than one in a million."

We cannot allow for the effect of enthusiasm or eagerness to be impressive. A person whose knowledge of chemistry is "profound" could not possibly make this mistake. He would be too accustomed to the mechanics of dilution not to get closer to the truth than that.

The chemical nomenclature placed in Holmes's mouth by Conan Doyle is old-fashioned and at times downright wrong.

In "A Case of Identity," Watson questions Holmes concerning the mystery of a missing person. "Have you solved it?" he asks. Holmes, far more interested in a chemical investigation he is carrying on, answers, "Yes. It was the bisulphate of baryta."

What a chemist would have said, however, would have been "barium bisulphate" or even "barium acid sulphate." The compound has the formula $Ba(HSO_4)_2$ and is an obscure one of no importance. It is no more than mentioned (sometimes not even that) in sizable reference books and is not a particularly difficult substance to analyze. Working on it should in no way have impaired Holmes's concentration on the human mystery.

In "The Adventure of the Copper Beeches," the necessities of investigation *do* interfere with Holmes's chemistry. On learning that he must take a train at a certain time, Holmes says, "Then perhaps I had better postpone my analysis of the acetones . . ."

What can he be thinking of? Acetone is a specific chemical compound, $CH_3COCH_3$, and should not be used in the plural as though it represented a class of compounds. To be sure, it is the best-known member of a class known as the "ketones," a term derived from the German spelling of acetone. An amateur might therefore refer to the ketones as the acetones, but *not* a chemist of the caliber that Holmes is reputed to be.

In "The Adventure of the Engineer's Thumb," mention is made of counterfeiters who have been producing half-crowns made of some metal less valuable than silver. Holmes comments: "They are coiners on a large scale, and have used the machine to form the amalgam which has taken the place of silver."

Here we have another mistake. What has taken the place of silver is an "alloy," a term which refers to any mixture of metals. When the coiners' den has burned down "large masses of nickel and of tin were discovered stored in an outhouse." Presumably, then, the metal used for the counterfeit coins was a nickel-tin alloy.

Is it not possible to use the word "amalgam" as a synonym for "alloy" as Holmes did? To be sure, amalgam can be used to indicate not only a metal mixture, but a mixture of any kind whatsoever, but only nonchemists would do it. To a chemist such as Holmes, an amalgam is not only an alloy, but one particular variety of alloy. It is a mixture of mercury and any other metal. No true chemist would refer to any mixture not containing mercury as an "amalgam."

Or consider "The Adventure of the Blue Carbuncle."

A "carbuncle" is a precious stone that is a variety of garnet and is, chemically, an iron-aluminum silicate. It is deep red in color and it is to that it owes its name, for it has the color of a glowing bit of burning coal (from the Latin "carbunculus" for a "little piece of coal"). There are different varieties of garnet of different colors, but only the red ones are called carbuncles. Hence a "blue carbuncle" is a contradiction in terms.

In the course of the story, Holmes says, "There have been two murders, a vitriol-throwing, a suicide, and several robberies brought about for the sake of this forty-grain weight of crystallized charcoal."

Ignore the point that jewels are weighed in carats rather than grains, so that he should have referred to it as a "thirteen-carat weight."

Much more important is the fact that a carbuncle is not "crystallized charcoal." A carbuncle is a compound of iron, aluminum, silicon, and oxygen. Charcoal, on the other hand, is at least 90 percent carbon.

Holmes is confusing a carbuncle and a diamond. A diamond is indeed pure carbon and can be referred to as "crystallized charcoal," although a good chemist is much more likely to say "crystallized graphite" or "crystallized carbon."

We can see where it is possible to suppose carbuncles to contain carbon from the identity of the first syllable, but that is a coincidence that traces back to color and the Latin language. A chemist would simply never make this particular mistake.

Finally, there is the occasion in "The Sign of the Four" when Holmes decides to rest his mind by taking it off a case and spending some time on chemistry. He says, "When I had succeeded in dissolving the hydrocarbon which I was a work at, I came back to the problem of the Sholtos . . ."

Hydrocarbons are composed of molecules made up of carbon atoms and hydrogen atoms only. Those with large molecules are soft solids at ordinary temperatures (tar, pitch, asphalt); those with small molecules are liquids at ordinary temperatures (kerosene, gasoline, naphtha).

Hydrocarbons mix with each other freely. If a solid hydrocarbon is placed in a liquid hydrocarbon, the solid hydrocarbon will mix with and easily dissolve in the liquid. What we call "dry-cleaning" is an example of this. Some liquid hydrocarbon (or chemically similar substance) succeeds in dissolving stains out of textile material because those stains are sufficiently closely related to hydrocarbons, in whole or in part, to dissolve easily in the liquid.

What Holmes is really saying, then, in connection with the hydrocarbon he wanted to dissolve was, "As soon as I had used my dry-cleaner . . ." That particular problem could not have succeeded in resting his brain for more than forty-five seconds.

Yet is there nothing to be said on the other side? Was Conan Doyle never prescient, even if only by accident?

Yes, he was. There is a remarkable passage in "The Adventure of the Devil's Foot." There Conan Doyle introduces an imaginary root ("devil's-foot root") obtained from West Africa. If this is ground to a powder and the powder set on fire, it produces a toxic smoke or fume that maddens and kills.

With more bravery than good sense, Holmes tests the substance on himself and on the ever-loyal Watson. Here is how Watson describes the effect:

I had hardly settled in my chair before I was conscious of a thick, musky odour, subtle and nauseous. At the very first whiff of it my brain and my imagination were beyond all control. A thick, black cloud swirled before my eyes, and my mind told me that in this cloud, unseen as yet, but about to spring out upon my appalled senses, lurked all that was vaguely horrible, all that was monstrous and inconceivably wicked in the universe. Vague shapes swirled and swam amid the dark cloud-bank, each a menace and a warning of something coming, the advent of some unspeakable dweller upon the threshold, whose very shadow would blast my soul. A freezing horror took possession of me. I felt that my hair was rising, that my eyes were protruding, that my mouth was opened, and my tongue like leather. The turmoil within my brain was such that something must surely snap. I tried to scream, and was vaguely aware of some hoarse croak which was my own voice, but distant and detached from myself."

A half-century later, the physiological effects of lysergic acid diethylamide (LSD) were discovered — though not in an African root — and the effects were not very different from those Watson described. It seems that Holmes and Watson had the equivalent of a "bad trip" decades before its time.

This is a remarkable bit of chemical science-fiction that came true, and it makes up to me for all the bits of poor chemistry Conan Doyle inserted into his stories.

# Part V

# Science: Explanation

# 28

# The Global Jigsaw

I suspect that many schoolchildren, in poring over the map of the world during their geography classes, have noticed that the eastern coast of South America rather resembles the western coast of Africa. I don't suppose many children actually defaced the map to check on the matter, but if they had cut out both South America and Africa they would have found that the bulge of Brazil fit neatly into the Cameroon coastline, like two parts of a jigsaw puzzle.

As a matter of fact, this coincidence in shape was noted as soon as the two continental coastlines were mapped with reasonable accuracy. The English scholar Francis Bacon pointed it out in 1620.

But was it a coincidence? Could it be instead that Africa and South America were once joined, that they split apart along the line of the present coasts and then drifted apart?

The first person to deal thoroughly with this notion of "continental drift" was a German geologist, Alfred Lothar Wegener, whose passion in life lay in the exploration of Greenland. In 1912, he published *The Origin of Continents and Oceans,* in which he suggested, essentially, that the continents floated slowly across the surface of the earth.

The continents, which are chiefly granite, are less dense than the rocks of the ocean floor, which are chiefly basalt. This is why the continental blocks ride high and lift themselves above sea level. Slowly, he said, they would drift this way and that.

Wegener felt that originally all the continents existed as a single vast block of land set in a single vast ocean. The supercontinent he called "Pangaea" (from Greek words meaning "all earth.") For some reason,

135

Pangaea broke into fragments and the fragments drifted apart, forming the present land masses as separate parts of a global jigsaw.

There was a great deal that was attractive about the hypothesis. The continents did seem to fit together, especially if the continental shelves were considered as their edges rather than the actual coastlines. Nor was it just a matching of shapes; there was a geologic fitting of the nature of the coastal rocks as well.

What's more, such continental drift might be the answer to a biological puzzle. There are similar species of plants and animals that exist in widely separated portions of the world; portions separated by oceans that those plants and animals could surely not have crossed.

In 1880, the Austrian geologist Edward Seuss had explained this by supposing there had once been land-bridges connecting the continents. Large tracts of land had risen and fallen, he said, serving as land-bridges in one epoch and as sea-bottom in another.

It seemed neater, somehow, to suppose that plants and animals had evolved and spread over Pangaea and that when the supercontinent split up and the parts drifted away, similar species trapped on those drifting parts slowly came to be separated from one another by thousands of miles.

Finally, while fossils located in the sedimentary rock of the continents (where bogs, lakes, or shallow estuaries had once existed) were up to 600 million years old, fossils from the Atlantic sea-bottom were much younger — as though the Atlantic Ocean itself were much younger than the adjoining continents.

Yet none of these points made Wegener's theory of continental drift acceptable to geologists. The theory was derided by some and ignored by others, and when Wegener froze to death in Greenland in the winter of 1930 during his fourth expedition to the Arctic island, his notions still seemed to amount to nothing of value.

It was not that geologists were being closed-minded and reactionary or that they were displaying a curmudgeonly refusal to see the obvious. There was a flaw in Wegener's theory that was a ruinous one. The continents simply could not and therefore did not float on the basalt beneath. The basalt was too stiff and firm.

The final evidence in that connection came in 1958, when the first American satellite, Vanguard I, was launched. It showed the earth's shape to be a bit uneven, bulging slightly here and depressing slightly there. In order to maintain that uneven shape against the pull of gravity, the rocks immediately below the surface had to be stiffer than steel.

It was therefore completely impossible for the continents to float and, however neatly Wegener's theory explained a dozen puzzles and coincidences, it could not be accepted.

In 1960, therefore, when I published the first edition of my book *The*

*Intelligent Man's Guide to Science,* I gave Wegener's theory just one paragraph and said, "The theory eventually foundered on hard facts."

But, even as those "hard facts" were established, a whole new set of facts just as hard were being uncovered.

At the time Wegener wrote his book, the nature of the ocean bottom was almost totally unknown. A few soundings here and there had been made by heaving a plumb line overboard, but that amounted to just about nothing.

During World War I, however, methods for estimating distance by means of ultrasonic echoes from objects underwater (now called "sonar") were worked out by the French physicist Paul Langevin. In the 1920s, a German oceanographic vessel began to make soundings in the Atlantic Ocean by sonar and, by 1925, it was shown that a vast undersea mountain-range wound down the center of the Atlantic Ocean through all its length. Eventually, this was shown to wind through the other oceans as well and, indeed, to encircle the globe in a long, serpentine, "Mid-Oceanic Ridge."

After World War II, the American geologists William Maurice Ewing and Bruce Charles Heezen tackled the matter, and by 1953 they were able to show that running down the length of the ridge, right down its long axis, was a deep canyon. This was eventually found to exist in all portions of the Mid-Oceanic Ridge, so that it is sometimes called "The Great Global Rift."

The Great Global Rift divides the earth's crust into large plates that are, in some cases, thousands of kilometers across. These are called "tectonic plates," from the Greek word for "carpenter," since the various plates seem so neatly joined together.

The joints are not always in the mid-ocean regions. One joint skims the borders of the Pacific Ocean, cutting across the western coast of California. The famous San Andreas fault is part of that joint. Another joint runs up through Eastern Africa along the long, narrow lakes of the region, then through the Red Sea and up the Jordan River Valley.

There seems a clear connection between the locations of the joints and the tendency for some regions to experience earthquakes and volcanic eruptions. The joints are clearly not quiet places.

What's more, the Great Global Rift seemed to be volcanic itself. In 1960, only two years after Vanguard I seemed to have killed continental drift once and for all, the American geologist Harry Hammond Hess presented evidence in favor of "sea-floor spreading." Hot molten-rock slowly wells up from great depths into the rift in the mid-Atlantic, for instance, and solidifies at or near the surface. This upwelling of solidifying rock forces the two plates on each side apart. As the plates move apart, South America and Africa are forced apart.

In other words, the continents may not drift, but they may be pushed.

This new material made me not so certain that the continents did not change position. In the second edition of my *Guide to Science,* published in 1965, I no longer said that continental drift had "foundered on hard facts." I said, more cautiously, it had to "face some hard facts."

Evidence for sea-floor spreading rapidly grew more impressive. If it were true, the floor of the Atlantic Ocean ought to be oldest at its edges and younger as one approached the Great Global Rift from either side. Every method of judging the age of the sea-floor supported that.

It seems that the direction of the earth's magnetic field periodically shifts, and the Atlantic sea-floor shows a pattern of these shifts as one moves out from the rift, and does so symmetrically on each side.

By the late 1960s, sea-floor spreading seemed an incontestable fact. This did not re-establish Wegener's theory of continental *drift*, because drift was still impossible. A new mechanism, the shifting of the plates was established, however, and all the consequences of Wegener's theory fell into place.

The scientific defenses at once went down, as they should have. The shifting of continents, the existence of Pangaea 225 million years ago, and its split-up — all this was accepted as readily as they had previously been opposed rigidly. In fact, the theory was found to account so elegantly for volcanoes, earthquakes, island chains, ocean deeps, mountain ranges, and many evolutionary facts, that it quickly became the central dogma of geology.

In the third edition of *Guide to Science,* published in 1970, I described what was now called "plate tectonics" in detail and referred ruefully to that paragraph in the first edition ten years before. It was all a remarkable example of the way in which science can not only advance, but can change its mind while doing so.

# 29

# The Inconstant Sun

We live by grace of the sun. All of life on Earth is the gift of the sun.

It is no secret. Human beings knew it long before they had developed what we call civilization. In north temperate latitudes, the sun was watched anxiously, not merely for the sunrise that would yield light and at least a measure of warmth at the end of the long, cold, dark night, but also during the months of its decline, which marked the inevitable coming of winter.

Through all the summer and fall, through all the days of dwindling warmth and increasing chill, the noonday sun attained each day a lower height in the southern sky than the day before. It gave less heat each day, and there had to be the fear that, though it had not happened in previous years, in *this* year the sun would sink indefinitely, disappear beyond the southern horizon, and leave the world to darkness, cold, and death.

It never happened. The sun sank at a steadily lower rate and the day came when it reached the lowest spot past which it would not sink. We call that the "winter solstice" ("standstill of the sun"). It falls on the day we now call December 21, and though the bitter winter lay ahead, the noonday sun was climbing higher and higher and the promise of eventual spring and rebirth was sure.

The rise should not continue indefinitely either, for otherwise there would be increasing heat and drought until life became impossible. Always, however, there was an upper limit, too, to the sun's position, the "summer solstice," which we now mark as falling on June 21.

Ancient peoples the world over, long before they had writing or any but the simplest technology, worked out ways of keeping track of the shifting sun. Stonehenge, in southwestern England, that circle of enormous rocks, is the best-known of these ancient "observatories," where the position of the sun at each solstice could be marked by sighting along particular rocks at sunrise, where the turnabout, the beginning of the new increase or decrease in height, could be checked.

Naturally, the good behavior of the sun was the occasion for an outpouring of relief, for celebration and festival—particularly at the winter solstice, when death seems more imminent and, in fact, already present.

The Romans celebrated with a weeklong Saturnalia (Saturn was their god of agriculture, and the solstice meant that crops would eventually be grown again). It was a time of holiday, of joy, of merriment, of feasting, drinking, and the giving of gifts, a time to preach the brotherhood of man. The climax of the celebration came on December 25, which, in the days of the early empire, the Mithraists celebrated as "the day of the sun."

It was the happiest time of the year and the early Christians, unable to beat that bit of heathenism, adopted it. In the fourth century, they made December 25 the anniversary of the birth of Jesus, though there was no biblical warrant for that at all; and to this day we celebrate the winter solstice as "the day of the son," so to speak, and it is still a Saturnalia.

The feared death and the apparent rebirth of the sun, with the concomitant apparent death of the plant world in the winter and its rebirth in the spring, gave rise to myths concerning the death and resurrection of gods—of Osiris among the Egyptians, of Adonis or Tammuz in the Near East, of Persephone among the Greeks. Some of that lingers in the West today in the Good Friday/Easter celebration of death and resurrection. All these myths gave promise that human death, too, is but a passing phase, and that there would be a resurrection in a better world.

Thus, the shifting inconstant sun gave a powerful impetus to the minds of early human beings, pushing thinkers in the direction of astronomy, mathematics, and science generally, and in the direction of religion, too.

With the coming of civilization, with the development of writing and of record-keeping, the sun was tamed. Its wandering through the skies, up and down, no longer seemed willful and uncertain, but began to be seen as an endless, mechanical, and automatic cycle.

The Sumerians, who first developed writing and who lived in what is now Iraq, plotted the course of the sun along the constellations of the zodiac, and for centuries generations of astronomers worked out the details of its motion, along with those of the moon and the bright planets.

The sky lost its terrors, at least to the sophisticated, and the sun began to seem reliable and benevolent. To the medieval Christians, the sun was a lamp in the sky, a container of weightless light that illuminated and warmed the earth steadily, and whose motions were ordained by God merely in order to provide the seasons and to give humanity a way of developing a calendar and of marking the coming of the holy days. The sun, it was felt, would continue to perform its function without change until the final Day of Judgment when it would please God to put an end to it and to all the world.

In view of this and of the completely obvious dependence of all life upon the sun, it became easy to view the sun as the very symbol of the Godhead. It was round, brilliant, benevolent, reliable, unchanging, and in all ways perfect.

In 1609, however, came a blow that marked the end of the comfortable medieval view of the universe. In that year, the Italian scientist Galileo Galilei devised a telescope and turned it on the heavens. He discovered innumerable stars in the Milky Way, mountains and craters on the moon, four satellites of Jupiter, and so on.

He even found imperfection in the immaculate glory of the sun, for he detected *spots* on it.

To be sure, spots had been seen occasionally in earlier times. The sun, when setting, has its light sufficiently dimmed to enable people to look at it without harm, and occasionally it can be seen dimly through mist and again can be observed. At these times, a dark spot or two were sometimes reported, for the really large spots can be seen without a telescope.

Such occasional reports could be dismissed, however, as optical illusions.

Galileo, however, saw numerous spots, studied them carefully, and made drawings of his observations. He followed the spots from day to day, showing how they progressed across the face of the sun and were foreshortened near the edges. He reasoned that they were part of the surface structure of the sun and pointed out that their motion demonstrated that the sun rotated on its axis in 27 days. Others at that time who built telescopes of their own, once Galileo had shown the way, confirmed these findings in every detail.

Christian leaders, after some resistance, were forced to reconcile themselves to the imperfection of the sun.

Part of the imperfection lay in the fact that the spots seemed to appear on the sun's surface randomly, but, in time, order was imposed on that, too.

In 1825, a German pharmacist, Heinrich Samuel Schwabe, who had an amateur's interest in astronomy, along with a small two-inch telescope, took to watching the sun, since the exigencies of his business made it impossible for him to watch the skies at night. For seventeen years (!) he observed the sun on every day that it was visible, and he sketched the spots he saw.

In 1843, he was ready to announce that the sun grew spottier and spottier, reaching a maximum, and then growing less and less spotty until it was virtually spotless. It would then begin a new cycle. Each cycle, it turned out, lasted an average of 10.7 years.

At first, no one paid attention to Schwabe (a mere amateur) but, in 1851, the important scientist Friedrich Wilhelm von Humboldt, mentioned Schwabe and his findings in his encyclopedic summary of science, and with that began the modern era of solar astronomy.

But did it matter, after all, if the sun were spotty? Except for rare occasions, only astronomers could see the spots, and those spots didn't seem to affect the constant shining of the sun or its constant light and warmth. If the spots didn't affect the earth and the human beings who lived on it, who cared about them? (Except astronomers, of course, but, then, who cared about *them?)*

And yet the sunspots *are* important — to everyone on the earth!

The earth has a magnetic field. This has been known since 1600, and we make use of it. The mariner's compass depends upon it, and for centuries long-distance navigation depended on that. In 1852, a British physicist, Sir Edward Sabine, showed that the earth's magnetic field varied in intensity in a regular way, rising, then falling, then rising again, over the years.

Since the sunspot cycle had just been announced, it seemed reasonable to try to compare the rise and fall of the earth's magnetic field with the rise and fall in sunspots and, behold, they matched!

Since then, the sunspot cycle has become a very popular way of explaining cycles on earth. People have matched intensity of rainfall to the sunspot cycle, and through that, other things, too. Naturally, with the rise and fall or precipitation, you had a cycle of good crops and poor, prosperity and depression, feast and famine, optimism and suicide.

The penchant for finding cycles undoubtedly went far beyond what could be authenticated, but one had to ask the question: How could the rise and fall in the spottiness of the sun affect the earth in any way — even in the completely accepted form of an influence on the earth's magnetic field and on the auroras in the polar regions?

In time, the use of "spectroscopy," the careful analysis of light from the sun, and the observation of which wavelengths represented the peak of light-emission, made it possible to measure the temperature of the sun's surface. The unspotted portion of the surface had a temperature of about 6,000°C, while the spots were only 4,000°C. (The spots seem dark precisely because they are cooler than the surrounding unspotted, hotter, and therefore more blazingly brilliant areas.)

Could it be, then, that the overall temperature of the sun when it was particularly spotty was noticeably lower than when it was unspotted and that this had its affect on earth? Could it be that the sun is *not* constant and unvarying, and that the earth is exposed to a slow and shallow swing of hot and cold?

The answer would seem to be, after a fashion, yes, but oddly enough the spotted sun doesn't act as though it is cool. The earth seems to be more affected by a spotted sun than an unspotted one, and it should certainly seem that it would be a hotter sun that would do more to the earth than a cooler one would. How can the cool spots make the sun hotter?

The beginning of an answer came in 1859, when an English astronomer, Richard Christopher Carrington, noticed a starlike point of light burst out on the sun's surface, last five minutes, and subside. This was the first observation of a "solar flare." It is the opposite of a sunspot in many ways. Whereas a spot is a long-lasting region that is cooler than the sun's surface generally, the flare is a short-lived event that is hotter than the sun's surface generally.

The flares are somehow associated with the spots. (We don't know yet exactly how — but then we don't know exactly what causes the spots, or why the spots cycle as they do.) The spottier the sun, the more likely that flares will burst out here and there, and it is the flares that seem to affect the earth particularly.

When the sun is spotty, we therefore speak of an "active sun"; and only a couple of years ago such an active sun affected the earth in a most unusual and very direct way.

In 1973, the United States put a space-station, called "Skylab," into orbit. It was occupied by astronauts on three separate occasions, and it was thought that it would stay in orbit for about ten years. By that time, it was supposed, a space-shuttle would have been developed that could nudge Skylab into a higher orbit where it could remain indefinitely.

Unfortunately, the sunspot cycle reached its peak earlier and more intensely than had been expected. The sun was very active indeed and, thanks to flares and other turbulences on its surface, more energy than had been calculated was delivered to the earth's upper atmosphere.

This expanded the thin gases of the upper atmosphere, which bellied outward, so that Skylab, in its orbiting about the earth, passed through a layer of gas that was not quite as rarefied as had been assumed. Skylab lost energy of revolution faster than expected and was ready to descend to the earth after only six years. Unfortunately, numerous delays prevented the space-shuttle from being ready for service in time to be of help and there was nothing that could be done to keep Skylab from coming down. It might conceivably have done damage in its descent, but the earth is a huge target and the remnants that survived passage through the atmosphere landed, on July 11, 1979, in the Indian Ocean and western Australia. No damage was done, but if it had been, solar activity would have been one of the factors responsible.

What form does this influence of the sun on the earth take? Is it simply light and heat? How could light and heat affect the earth's magnetic field and its aurorae?

Actually, there is something else. The sun's heat and its turbulent activities throw matter upward in gigantic tornadolike storms. There are prominences, vast gouts of white-hot hydrogen tossed up in enormous towers that are visible at the edge of the sun during total eclipses, and even at other times when special equipment is used. As a result some matter is lost from the sun permanently and speeds away through the solar system in the form of a thin spray of gas. This gas is at such enormous temperatures, 1,000,000°C and more, that it does not consist of intact atoms. The electrons are stripped away and the bare atomic nuclei are exposed. Most of the gas is hydrogen, and hydrogen nuclei consist of single protons.

It follows, then, that streaming outward from the sun in all directions is a thin drizzle of electrons (each carrying a negative electric charge) and protons (each carrying a positive electric charge). A British physicist, Edward Arthur Milne, predicted in the 1920s that such an outward spray of matter was possible. Soon after World War II, rocket experiments by an Italian-American physicist, Bruno Benedetto Rossi, showed that the spray actually existed. In 1962, the American physicist Eugene Newman Parker referred to this as the "solar wind."

The solar wind spreads outward and reaches the orbit of the earth, and moves far beyond it, too. The earth, in other words, can be viewed as actually orbiting within the very, very thin outer reaches of the sun's atmosphere.

The solar wind does not actually hit the earth's surface. Its electrically charged particles are deflected by the earth's magnetic field. They spiral about the "lines of force" of that field, forming a huge doughnut of charge about the earth. This was detected in 1958 by rocket experiments supervised by the American physicist James Alfred Van Allen, and these radiation zones were therefore called the "Van Allen belts" at first. Nowadays, the region is referred to as the "magnetosphere."

The earth's magnetic lines of force curve down to the magnetic poles in the polar regions, and the magnetosphere curves down with it. Floods of charged particles enter the atmosphere at those points. They strike the thinly spread out atoms of the upper atmosphere and the energy of interaction is converted into light. The result is the auroras that are almost continuously visible in the polar regions.

The energetic solar flares with their enormous temperatures send gouts of matter upward in amounts far beyond those sent up by the normal surface of the sun. The eruption of a flare does not increase the solar wind generally, but produces a local solar "blizzard," so to speak, immediately above it.

Generally, such blizzards miss the earth, but every once in a while a flare sends out a stream of particles in the earth's direction and, after two days or a little more, it strikes the magnetosphere, flooding it and sliding down in wholesale quantities into the polar atmosphere, setting up auroras that are visible far beyond the usual latitudes.

The flood of charged particles produces a "magnetic storm" that does not affect human beings on the earth's surface in any ordinary way, but produces disruptions in modern electronic technology.

An example of this took place in 1944, when Great Britain's radar network suddenly went completely out of whack. For a horrifying period of time, the British and their allies thought the Nazis had worked out a way of countering radar defenses, but then it turned out that a giant flare was responsible. Gradually, the charged particles were dissipated and the radar returned to normal.

Nowadays, our growing dependence on electronic communications and controls of all sorts has made us continually more subject to disruption by events 93 million miles away on the sun. Such disruptions might even include the disabling of the sophisticated controls over our missiles, for instance, or our ability to detect and respond to enemy attacks. (To be sure, the enemy might also be disabled.) These dangers—which include radiation risk to astronauts in space—are more likely, the more spotted the sun, so that we have new reason to be cautious in those years when the sun is active.

So, of course, we must view the sun with more respect. However much its light and heat may seem constant, its activity varies unpredictably, with important effects. The sunspot cycle may seem a regular phenomenon, but it isn't quite. Peaks of spottiness may be separated by as little as 7 or as much as 17 years, and one peak may be two or three times as high as another.

In fact, the situation is worse than that. In 1893, the British astronomer Edward Walter Maunder, checking through early reports in order to gather data for the sunspot cycle prior to Schwabe's time, was astonished to find that there were virtually no reports on sunspots between the years 1645 and 1715.

Galileo and other astronomers had reported numerous sunspots between 1609 and 1645, but then the reports stopped. It wasn't that nobody looked. There were astonomers in the late 1600s, competent professionals and good observers, who reported searching for spots and failing to find them.

Maunder published his findings in 1894, and again in 1922, but no one paid any attention to him. By that time, the existence of the sunspot cycle was well-established and astronomers were as reluctant to believe in an unspotted sun in 1900 as they would have been reluctant to believe in a spotted one in 1600.

But then, in the 1970s, the American astronomer John Eddy came across Maunder's work and decided to check it. To his own surprise, he found that Maunder's report was accurate. Eddy even went beyond

Maunder, looking up reports of naked-eye sightings of sunspots in records going back to the fourth century B.C., both in Europe and in the Far East. He found that every once in a while there were periods of many decades during which there were no sightings. Apparently, every once in a while the sun underwent a "Maunder minimum," during which it remained virtually unspotted for a long period of time and after which it returned to the normal situation of the sunspot cycle. The period from 1645 to 1715 was merely the most recent occasion.

Eddy did not let it go at that. He checked the matter in ways Maunder did not—for lack of information. Eddy knew that auroras were more numerous and intense at times of high-spottiness of the sun, and it would only be then that they would be visible in the latitudes of London and Paris. There should be occasional auroras visible during the sunspot cycle; none during a Maunder minimum. There were many reports of auroras after 1715, quite a few before 1645; none between 1645 and 1715.

The corona, visible about the sun during a total eclipse, has one shape when the sun is active and another when it is inactive. During the Maunder minimum, all descriptions of the corona were those of an inactive sun.

Finally, cosmic rays entering the earth's atmosphere form radioactive carbon-14 in small (but easily detectable) quantities. These are absorbed by plants and can be detected in wood. When the sun is active, its magnetic field expands and protects the earth from cosmic rays to a certain extent, and less carbon-14 is formed. During a Maunder minimum, this protection is absent for a long time and more carbon-14 is formed. If tree rings are analyzed for carbon-14 content, it turns out that the content is high throughout the years of the Maunder minimum.

It would seem, then, that there is no doubt that the sun is more complicated than we think. The sunspot cycle is itself part of a larger cycle in which the sunspot cycle exists and is absent, alternately. A Maunder minimum can last anywhere from 50 to 200 years. Prior to the last there was one from 1400 to 1510, and prior to that one from 1100 to 1300, though the earlier ones aren't as certain as the last. There may have been twelve Maunder minima during historic times.

Why does the sun behave in this way? No one knows. When will the next Maunder minimum come? No one knows.

Do Maunder minima have any effect on the earth and on humanity? Perhaps. An unspotted sun delivers somewhat less energy to the earth than a spotted one. The earth should cool off a little when the sun is unspotted. However, the sun doesn't stay unspotted long. A new cycle of spottiness starts as soon as the old one is over, so that the coolness is temporary and unimportant—except during a Maunder minimum. During a Maunder minimum, the coolness accumulates and after several decades it should become noticeable.

This is no fun. People might welcome a reduction in heat-waves, but a general coolness cuts the growing season from a few days to a few weeks, lowers the amount of grain harvested, and increases the risk of famine. As a matter of fact, the period from 1645 to 1715 includes what historians sometimes call the "little Ice Age," when times were hard and famine stalked Europe. The winter of 1709-1710 (when the War of Spanish Succession was raging) was a time of record cold, and the suffering in France, which was losing the war, was immense and pitiable.

In the earlier Maunder minimum, from 1400 to 1510, the situation in Greenland, very bad at best, grew worse, and the Viking settlement that had hung grimly on for four and a half centuries was finally wiped out, thanks in part to the total impossibility of growing crops.

With that in mind, the question of when the next Maunder minimum is due has a new urgency; and the fact that we do not know, presents a new grimness.

We experience periodic *real* Ice Ages, when the glaciers extend downward as far as New York City. These arise in part because the earth's orbit about the sun is not truly unchanging. Its eccentricity grows slightly larger and smaller in a certain period; so does the extent of the tipping of the axis. Similarly, the direction of the tipping of the axis goes slowly through a complete circle. When all these changes are at a certain crucial value, the earth gets a somewhat smaller amount of energy from the sun year after year and the glaciers are triggered into an expansion.

Suppose that the crucial value is not *quite* reached. A Maunder minimum at just that time might make things just enough worse to create the trigger. We can't be sure; we don't know enough; but perhaps—

Our uncertainties are even greater. The earth's magnetic field deflects the solar wind and shunts it into the atmosphere of the polar regions where, as it happens, only a very small percentage of the human population is to be found.

The magnetic field, however, at irregular intervals, diminishes in intensity to zero and then slowly begins to intensify in the opposite direction. Such "magnetic field reversals" have taken place dozens of times in the earth's history. The magnetic field is diminishing now and at a rate that will bring it to zero in about 1,500 years or so. For a few centuries it will remain low while it is building up under conditions that will have the magnetic compass pointing south.

What will life be like under such conditions? The solar wind will reach the atmosphere everywhere, more or less equally. Does that mean all latitudes will experience faint auroras after sunset and before dawn, with occasional stronger ones when flares strike? Will it mean that magnetic storms and electronic disruption will become commonplace? Will our climate be affected by the entry of charged particles from the sun into all portions of our atmosphere?

We can't predict. Such a thing has not happened before in the history of *Homo sapiens,* and we lack the data to come to reliable conclusions.

A second mystery even more puzzling involves certain tiny, nearly indetectable particles called "neutrinos."

The region of the sun that is of basic importance is the very core. It is there that the temperatures are up to 15,000,000°C and that pressures are equally enormous. It is there, under those temperatures and pressures, that hydrogen undergoes fusion and produces the energy that has kept the sun shining for 4.6 billion years.

What are the details of the processes that keep the fusion going within the complex structure of that core? How do these processes explain the sun's steady glow, the appearance of sunspots in a rising and falling cycle, the disappearance of sunspots in Maunder minima, the production of flares, and so on.

Astrophysicists have worked out what must be going on in the sun's core on the basis of the subatomic processes they have studied in the laboratory and the theories concerning them that they have evolved, but how can they possibly look into the core to check their conclusions?

There seems only one way. Some of the processes that are thought to take place at the core involve the production of neutrinos. Neutrinos differ from all other particles in that they pass through enormous thicknesses of matter as though they weren't there. Any neutrinos produced at the sun's core dart away at the speed of light, reaching the sun's surface in less than three seconds and reaching the earth (if they are pointed in the right direction) in eight minutes.

The neutrinos are hard to detect, since so few of them interact with atoms of matter—and unless they do, they cannot be detected at all. Nevertheless, the job can just barely be done, and scientists have worked for years with large detecting devices deep in mine shafts where no other form of radiation can penetrate.

Now comes the problem. The number of neutrinos detected is considerably less than theory has led the experimenters to expect. At the very most, the neutrinos seem to be emerging from the sun's core in only one-third the expected number. The detectors have been checked and seem reliable; astronomical theories have been checked and seem unshakable, but one or the other must be wrong.

There are at least three different types of neutrinos, however, and the detectors can only detect one kind—the kind the sun is supposed to produce. There is just a chance that neutrinos can shift their identities and that, though only one kind is produced by the sun, by the time it reaches the earth it has become a mixture of the three kinds so that only one-third of them are detected. —But that is, so far, *just* a chance.

If that is *not* so, then the only other alternative is that what is going on in the core of the sun is not what scientists think is going on — and yet nothing else seems to be possible.

Scientists have considered some pretty far-out possibilities to explain "the mystery of the missing neutrinos." Perhaps the most far-out is that whatever goes on in the core to produce the sun's energy has failed periodically in the course of the sun's history, for one reason or another, and that the sun has entered one of those periods of failure now. The energy produced at the sun's core takes a million years or so to reach the surface, but now an expanding shell of less energy may be working its way toward the surface. At some time in the future, the sun will suddenly dim and "go out" for an unknown period of time and the earth will freeze to death.

*This is not at all likely,* but it is a measure of the desperation of scientists that they must consider such a thing. Undoubtedly the mystery will be explained in a far less radical way, but that will require further neutrino-detection with better instruments, and perhaps further advances in nuclear physics.

Even yet, we are not through with the new puzzles that have arisen.

John Eddy, the astronomer who confirmed the Maunder minima, has been searching through the records kept for the apparent diameter of the sun and has come to the conclusion that it must be shrinking. If he is right, the sun is very slowly collapsing and every hundred years has a diameter about 870 miles less than the century before.

If the sun were to continue shrinking at this rate, then it would shrink to nothing in about 100,000 years.

There is no chance of the sun shrinking to nothing, of course (and, indeed, some astronomers searching other types of records than those Eddy has worked with, claim there is no shrinkage at all). There is a chance, though, that the sun undergoes a slow pulsation — a limited shrinkage, followed by a limited expansion, then a shrinkage again, and so on.

If so, how far does the sun shrink and expand, how long does a single cycle of shrinkage and expansion take, and what is the effect on the earth? We don't know, of course.

There may well be little or no effect. The amount of shrinkage and expansion may be very small. Then, too, as shrinkage takes place, the sun has a smaller surface and should deliver less heat; but shrinkage warms the sun so that it delivers more heat per unit of surface, and this tends to neutralize the effect of the smaller surface. A similar argument, in reverse, is true for the case of expansion. Still, we don't really know.

To summarize, then, humanity depends on the absolute reliability of the sun. Even a tiny hiccup (on the solar scale) — a small irregularity in radiation, a small increase or decrease, a small abnormality in the sunspot process or in any other aspect of the solar machinery—could have disastrous consequences for us.

And in recent years, we have learned that the sun is a far more complex heat engine than we had thought, that there is far more to it than we had expected, and that there are far greater chances of irregularity than we had dreamed.

It is enough to make us quite nervous!

# 30

# The Sky of the Satellites

Thanks to the Voyager space probes we have learned a great deal about the four large Galilean satellites* of Jupiter: the volcanoes of Io, the cracked glacier that covers Europa, the frozen craters of Ganymede and Callisto.

Yet if we could imagine ourselves on the surface of any of these worlds and somehow protected from the harsh conditions, it would probably not be the surfaces that would hold our attention most, not the volcanoes, not the craters, not the cracked glaciers. It would be the skies.

Consider Callisto, the farthest of the Galileans, 1,171,000 miles from Jupiter's center. Thanks to Jupiter's tidal effect, Callisto, like the other Galileans, faces one side always to Jupiter, as our moon faces one side always to us. This means that Callisto rotates on its axis, relative to the universe generally, in the same time it revolves about Jupiter — once in 16.69 days.

Standing on Callisto, we would see the sun make one complete circle in that time, which means it would move considerably more slowly in Callisto's sky than in our own. On Callisto, the time from sunrise to sunset would be some 200 hours rather than the average of 12 hours that the sun stays in Earth's sky.

The three Galilean satellites that are closer to Jupiter than Callisto is revolve about Jupiter more quickly, hence rotate more quickly, and therefore see the sun move more rapidly across the sky. On Ganymede, the sun moves from rising to setting in 84 hours; on Europa, in 42.7 hours; on

---

*They were first seen by Galileo in 1610.

Io, in 21.2 hours. However, even on Io, the closest of the Galileans to Jupiter, the sun moves across the sky at only a little over half the speed it moves across our own sky.

It would be a small sun, too, only 6 minutes of arc across, as compared with the width of 32 minutes of arc for the sun as seen from Earth. If enough of the sun's light were blocked so that it could be looked at, it would be seen, from Callisto, as just barely large enough to show a disc. This would be true from all other satellites of Jupiter and from Jupiter itself. The total light and heat the sun would give off, to any point in the Jovian system, would be only one-twenty-fifth what it gives off as seen from Earth.

If the sun, shrunken and small, does not bulk large in Callisto's sky, there is, however, something else—Jupiter.

Since Callisto faces only one side to Jupiter at all times, the satellite does not rotate at all relative to Jupiter, and the planet does not appear to move in Callisto's sky. If you are located anywhere on Callisto on the side facing Jupiter, then Jupiter will be in a particular spot in the sky and will stay there, day after day, year after year.

If you are in the very center of the side facing Jupiter, then Jupiter is directly overhead in the sky. If you move away from that center, then Jupiter will move in the sky in the direction opposite to that in which you are moving. If you move far enough, Jupiter will move to the horizon and, eventually, set. You will have moved over to the side that faces away from Jupiter.

If you are anywhere on the side of Callisto facing away from Jupiter, you will *always* be facing away; Jupiter will *never* be in the sky.

This all-or-nothing situation, Jupiter always in the sky or never in the sky, is true for the other Galilean satellites as well.

Callisto is more than four times as far from Jupiter as we are from our moon, but Jupiter is a giant. It would take 41 moons, side by side, to stretch across a width equal to the diameter of Jupiter. Consequently, despite Callisto's greater distance from Jupiter, Jupiter appears considerably larger in Callisto's sky than the moon does in ours. The apparent width of Jupiter in Callisto's sky is 4.3 degrees, which makes it 8.3 times the apparent width of the moon in our own sky.

Nor ought we to compare Callisto's Jupiter with Earth's moon on the basis of width alone. Callisto's Jupiter is not only larger than the moon from left to right, but from top to bottom as well. Jupiter's area, as seen from Callisto, is therefore nearly 70 times the area of the moon as seen from Earth.

The other three Galileans are closer to Jupiter and the planet therefore bulks correspondingly larger in their skies. From Ganymede, Jupiter has an apparent area nearly 200 times that of our moon; from Europa, about 625 times; from Io, nearly 1,500 times.

Size in itself would make Jupiter incredibly impressive as seen from Callisto, let alone as seen from the other Galileans, but it is not just a larger object we would be watching.

Our moon is simply a circle of calm silvery light (when it is full) with a few shadowy splotches that never change. Jupiter is striped in orange, yellow, and brown and would slowly exhibit changes as the Red Spot and other lesser objects, ever altering in fine detail, crossed from one side to the other in a period of 5 hours.

From Callisto, the stripes and the changes would be vague and hard to see with the naked eye, but from each successively closer satellite, the markings would be clearer, the changes more visible. As seen from Io, the giant globe of Jupiter would be an ever-changing kaleidoscope in visible rotation.

Jupiter would be bright, too, brighter than the moon as we see it. To be sure, Jupiter would not appear quite as bright as one would expect from its apparent size, for it is bathed in much weaker sunlight than our moon is. Countering this, in part, is the fact that the bright clouds of Jupiter reflect about 7 times as much of the light they do receive as the dark, bare rock of the moon's surface does.

Taking everything into account, when Jupiter is full in the sky of Callisto it shines 12.5 times as brightly as Earth's full moon does. Jupiter shines upon Ganymede with 35 times the light of our full moon; upon Europa with 85 times that light; and upon Io with 220 times that light.

Who could look upon anything else but Jupiter, even in the sky of Callisto, let alone that of Io?

Still, we mustn't be fooled into thinking that Jupiter outshines everything else. The sun, shrunken and dim though it is by Earth's standards, still outshines swollen Jupiter. When the sun is in Callisto's sky, it is 1,360 times as bright as Jupiter at its brightest. Even in Io's sky, the sun is 77 times as bright as Jupiter at its brightest.

Jupiter, like our moon, shines only by reflecting the light of the sun. Only half its globe is lit by the sun and, depending on where the sun is in relation to Jupiter and oneself, the face of Jupiter that is visible may be entirely light, entirely dark, or partly light and partly dark.

In other words, Jupiter will show phases when seen from its satellites, in just the same fashion and the same order that our moon does when seen from Earth (or, for that matter, that Earth does when seen from the moon.)

The moon goes through a complete cycle of its phases in a single revolution about Earth, or once every 29.5 days. The Galilean satellites circle Jupiter much more quickly than our moon circles Earth, since Jupiter's gravitational field is much more intense than Earth's. Therefore Jupiter goes through its phases completely in 16.7 days as seen from Callisto, in 7.16 days as seen from Ganymede, in 3.55 days as seen from Europa, and in only 1.77 days as seen from Io.

As the sun rises on any of these satellites, Jupiter (if we imagine it to be directly overhead) is a semicircle of light on its eastern side. The western semicircle, away from the sun, is dark. As the sun climbs in the sky, the lighted portion of Jupiter shrinks to a thick crescent, then to a thinner crescent. When the sun is high in the sky, it is the other side of Jupiter, the side hidden from us, that gets the light, and Jupiter is then only a dark circle in the sky.

As the sun continues to move toward setting, Jupiter begins to light up on the western side in a thin crescent that widens and widens until at sunset Jupiter is again a semicircle of light, on the side opposite to that at sunrise.

After the sun sets, Jupiter appears much brighter by the mere fact that the greater brilliance of the sun is no longer in the sky. What's more, however, light continues to spread over the face of Jupiter after sunset, and at midnight we have Jupiter at the full. It is then most magnificent as its circle of striped light against the black sky. (The sky is black in the daytime, too, on airless worlds.)

But time does not stand still. Once midnight is passed, the darkness begins to invade the western edge of Jupiter, farther and farther, until the light is back to a semicircle on the eastern half—and the sun rises again.

(Naturally, if you see the sun from a different portion of the satellite's surface, the pattern changes. If you place yourself so that you see Jupiter's sphere just topping the western horizon, it is in the full-phase at sunrise, in the new-phase at sunset and in the half-phase at noon and, again, at midnight.)

When the sun passes Jupiter in the sky of any of its satellites, it passes behind Jupiter and is eclipsed. Because Jupiter's axis is only slightly tipped, and because the satellites circle in Jupiter's equatorial plane, the eclipse takes place at every solar passage as seen from Io, Europa, or Ganymede. As seen from Callisto, Jupiter is comparatively small so that the sun occasionally misses the planet's globe, moving above or below it, avoiding an eclipse.

On Earth an eclipse of the sun by the moon lasts at most 7 minutes. As seen from the Galilean satellites, the sun is so small and Jupiter so large that the eclipse can last for hours.

On Io, the nearest of the satellites, an eclipse of the sun can last for up to 2.2 hours. As we move to the satellites farther from Jupiter, the planet becomes smaller in apparent size, but the sun moves more slowly across the sky and that more than makes up for it. Consequently, eclipses grow longer as we move away from Jupiter. They can be 2.8 hours long on Europa, 3.5 hours on Ganymede, and 4.6 hours on Callisto.

To be sure, an eclipse of the sun by Jupiter does not have the same effect as an eclipse of the sun by our moon does. Jupiter is so much larger than the sun (when both are seen from the Galilean satellites) that the Solar corona is completely covered. Even if the corona weren't covered it would be only one-twenty-fifth the size and brightness it is when seen from Earth.

Jupiter, however, is itself the sight to be seen. While the sun is in eclipse, Jupiter is a black circle in the sky.

Of course Jupiter is a black circle against a black sky, but the circle is still evident because it hides the stars behind it. Since there are no atmospheres to speak of on the Galilean satellites, there is nothing to absorb starlight and we can see perhaps twice as many stars from their surfaces that we can see from the surface of Earth.* What's more, the stars, as seen from the Galilean satellites, are sharper and do not twinkle. With the sun hidden and Jupiter dark, those unusually sharp pinpoints of light overspread the heavens quite thickly, except that none at all appear within the dark circle of Jupiter.

But there is much more to it. The light from the sun shines through and is scattered by the outermost reaches of Jupiter's atmosphere on all sides. As a result, a red-orange circle outlines the dark globe of Jupiter. (Actually, because of Jupiter's rapid rotation, it has an enormous equatorial bulge, and its outline, as seen from the satellites, is slightly elliptical rather than perfectly circular.)

If the sun is directly behind Jupiter's center, the gleam of Jupiter's atmosphere produces a uniformly bright circle. If the sun passes Jupiter somewhat off-center, or if the time is considerably before or after the middle of the eclipse, then the circle of light is unevenly brilliant and gleams more on one side than on the other.

All the satellites produce a spectacular eclipse show. On Callisto, it can last the longest, but on Io, where the maximum length is less than half as long as that on Callisto, the dark circle outlined by light in Io's sky has 18 times the area of the dark circle in Callisto's sky.

In addition to the sun and Jupiter, one can see in the sky of each of the Galilean satellites, the three *other* Galilean satellites. The three satellites that one sees in the sky, vary in apparent size and brightness depending on where they are in their orbits. They can pass behind Jupiter, or move into Jupiter's shadow and be eclipsed before they pass behind Jupiter's bulk, or after one expects them to be emerged.

From Callisto, the three other Galilean satellites, which are all nearer Jupiter, seem to hug Jupiter's globe as they wheel about it. From a portion of Callisto's surface where Jupiter is high in the sky, the three other Galileans move from one side of Jupiter to the other and back again, at different speeds, never rising or setting, and forming an ever-changing pattern that must act hypnotically on any watcher.

---

*They form the familiar constellations, though, and are a touch of home. The stars are so distant that shifting our position from Earth to Callisto makes no difference, and we see them just about the same from either vantage point.

If one stands on the point of Callisto's surface where Jupiter is precisely on the other side of the satellite (under your feet, so to speak), none of the other three Galilean satellites are ever seen in the sky either — only the stars at all times, and the sun half the time. As one moves on Callisto's surface to a point where Jupiter is just below the eastern or western horizon, Jupiter itself may never rise, but each of the three other Galileans takes its turn in rising, climbing some distance up in the sky, then turning and setting near the place of rising.

On the other satellites, the pattern is different. From Io, for instance, the other three satellites (all farther from Jupiter) all make a complete circle of the sky, rising, moving behind Jupiter, setting, then moving around Io and rising again. From a point on Io's surface directly away from Jupiter, the planet may never be seen in the sky, but each of the three other Galilean satellites rises, moves across the sky, and sets in the opposite quarter.

From Europa, Io is seen to hug Jupiter, while Ganymede and Callisto make the complete circle. From Ganymede, Io and Callisto hug Jupiter, while only Callisto makes the complete circle.

From each of the Galilean satellites, the panorama of the skies is so enormously impressive and so fascinating in its variety that returning to the bloated sun and the one pale moon of Earth's sky might well seem like an unbearable loss for which nothing could compensate.

# 31

# The Surprises of Pluto

Ever since Pluto was discovered, it has been described as the planet farthest from the sun. Its orbit makes a huge sweep that it takes Pluto 248 years to traverse, instead of the single year in which Earth completes one turn about its own small orbit.

Pluto's orbit, however, is distinctly elliptical, with the sun well to one side of the center. When Pluto is at the end of its orbit that is farthest from the sun, it is 4.6 billion miles away, and it is then 1.7 times as far away from the sun as is Neptune, the next farthest planet.

Every 248 years, however, Pluto moves around to that section of its orbit where it is closest to the sun, and then it is only 2.7 billion miles away. Surprisingly, it is at that time actually a trifle closer to the sun than Neptune is. For twenty years it skims along that portion of its orbit, remaining closer than Neptune. Then it passes beyond Neptune's orbit again and begins its long trek outward to the vast distances beyond.

In January 1979, Pluto passed inside Neptune's orbit. Pluto is therefore *not* the farthest planet from the sun; Neptune is, and will be until 1999, when Pluto will resume its position as farthest planet and won't pass inside Neptune's orbit again till 2227.

Pluto was first discovered because the outer planets, Uranus and Neptune, don't move quite exactly as the law of gravitation predicts. The difference is minute, but some astronomers wondered if there might be another planet beyond Neptune whose gravitational pull wasn't being allowed for. If the planet's pull *were* taken into account, that might explain the discrepancy in the motions of Uranus and Neptune.

About 1900, the astronomer Percival Lowell calculated where the distant planet should be to account for the discrepancy, and he looked for it. It wasn't an easy task. The planet would be so far away that it would be very dim and would be lost among many thousands of equally dim stars. When Lowell died in 1916, he still hadn't found it. His observatory continued the search and, in 1930, a young astronomer, Clyde William Tombaugh, finally located the planet.

He named it Pluto, after the god of the underworld, because it was so far from the light of the sun—and because the first two letters stood for Percival Lowell.

But there was a surprise. Pluto was considerably dimmer than had been expected. That meant it might be considerably smaller than had been expected. Instead of being much larger than Earth, as Uranus and Neptune were, it seemed to be only Earth-size at best. That was troublesome because if it were that small its gravitational pull wouldn't be enough to account for the discrepancy in the motions of Uranus and Neptune.

Pluto was so distant, though, that it was impossible to measure directly how large it was. The conclusion of smallness, just from its dimness, seemed uncertain.

Then, on April 28, 1965, Pluto was scheduled to pass very close to a certain faint star. The path of Pluto's center, which could be marked out very accurately, was going to miss the star by a small distance. If Pluto was as large as Earth, its surface would be far enough from its center to hide the star as Pluto passed it. In fact, the larger Pluto was, the longer the star would remain hidden.

A dozen excellent telescopes were trained on Pluto, and all registered the same surprising fact.

Pluto passed the star exactly on schedule, but the star kept right on shining, unconcerned. It wasn't hidden for even a fraction of a second.

Apparently Pluto was so small that its surface wasn't far enough from the center to reach the star. In order for this to be so, Pluto's diameter had to be less than 4,200 miles. Pluto was by no means as large as Earth; it was only as large as Mars, a planet with only half Earth's diameter and only a tenth of Earth's mass.

Even that isn't the end of the story.

Now that Pluto is about as close to the sun (and to Earth) as it ever gets, telescopes point toward it frequently. On June 22, 1978, an astronomer, James W. Christy, examined photographs of Pluto that he had taken and noticed a distinct bump on one side. He examined other photographs, and the bump was on those, too. What's more, the position of the bump changed.

It seemed that Pluto had a satellite, a smaller body that circled it. Christy named the satellite Charon, after the ferryman who takes the dead to Pluto's underworld kingdom.

Charon and Pluto are only 12,500 miles apart, roughly one-twentieth the distance that separates Earth from the moon. It takes Charon only 6.39 days to circle Pluto.

When you have two bodies, with one circling the other in a given time at a given distance, it is possible to calculate the combined mass. From the comparative brightness, the mass of each can be determined.

As a result, it now seems we have Pluto's real size. It is not as large as Earth—it is not even as large as Mars. In fact, the surprise is that it's not even as large as our moon. Pluto is only about 1,850 miles in diameter, as compared with our moon's diameter of 2,160 miles. And since it is very likely that Pluto is made of lighter material than the moon, astronomers estimate that Pluto is only one-eighth as massive as the moon.

Pluto is scarcely a respectable planet; it is more like a large asteroid.

Charon is smaller still of course. It is only 750 miles across and is only one-tenth as massive as Pluto.

Remember, though, that Lowell pointed out the approximate location of Pluto from the effect he thought it would have on the other outer planets. Pluto is so small, however, it couldn't possibly have any noticeable effect on them. The fact that Pluto was about where Lowell thought it should be was simply an extraordinary coincidence.

And that leaves us with a question: If it isn't Pluto that's affecting the other outer planets, what is? Is there another planet somewhere out there, a larger one, that we haven't discovered yet?

# 32

# Neutron Stars

Stars, like ourselves, are composed of atoms, and atoms are mostly empty space. At the center of each atom is a tiny "nucleus," and at the outskirts are very light "electrons." The nucleus contains a positive electric charge; the electrons contain a negative electric charge.

If anything acted to crush atoms together, forcing the electrons into the nucleus, then the opposite electric charges would cancel out. The whole atoms would turn into tiny uncharged "neutrons."

If the atoms of the whole of Earth collapsed into neutrons, all the matter of such a collapsed Earth would make up a sphere only 140 feet across. If the atoms of the sun collapsed into neutrons, there would be left a sphere only 8 miles across.

The only force that can bring about the collapse of a star is the mighty gravity of the star itself. What keeps the sun from collapsing under gravitational pull is the heat it develops from nuclear reactions at its center.

These nuclear reactions consume hydrogen, and billions of years hence, when the sun runs out of hydrogen, it will collapse. It won't collapse all the way to neutrons because it isn't quite large enough and its gravitational pull isn't quite strong enough. Stars larger than the sun might, however, collapse into tiny "neutron stars."

The theory of neutron stars was worked out in the 1930s, but is the theory correct? How can neutron stars be detected if they are only a few miles across and if they are thousands of billions of billions of miles away? It looked as though neutron stars would just remain speculations.

In the 1950s and 1960s, however, astronomers were studying radio waves coming from various portions of the sky. It seemed to them that some of the radio waves varied quite rapidly in intensity, almost as though they were twinkling.

At Cambridge University Observatory, Anthony Hewish devised a special "radio telescope" to study such twinkling. In July of that year, his student Jocelyn Bell detected very rapid bursts of radio waves in one particular part of the sky. They came with fantastic regularity, one burst every 1.33730109 seconds.

Hewish called it a "pulsating star," a phrase that was quickly shortened to "pulsar." Other pulsars were found and over a hundred of them are now known.

What could be sending out such rapid bursts so regularly? Some object had to be enormously massive to produce such energetic bursts, and it had to be spinning very rapidly to produce a burst every second. Ordinary stars were massive enough but were too large to spin that rapidly. They would fly apart if they did.

A neutron star, only a few miles across, could spin hundreds of times a second, however. Astronomers decided that pulsars had to be neutron stars. Nothing else would fit.

The most rapid bursts yet detected come from a vast cloud of gas called the "Crab Nebula." That cloud of gas is what is left of an enormous star-explosion that took place about a thousand years ago. Such explosions are exactly the sort of thing that would serve to collapse a large ordinary star into a tiny neutron star.

In the Crab Nebula is a pulsar that sends out a burst of radio waves every thirtieth of a second, and at the point where the radio waves originate is a dim star. In January 1969, that dim star was photographed in very short time-intervals and was found to go on and off thirty times a second. Even the light waves came out in bursts and a neutron star was finally seen!

# 33

# Black Holes

Of all the odd creatures in the astronomical zoo, the "black hole" is the oddest. To understand it, concentrate on gravity.

Every piece of matter produces a gravitational field. The larger the piece, the larger the field. What's more, the field grows more intense the closer you move to its center. If a large object is squeezed into a smaller volume, its surface is nearer its center and the gravitational pull on that surface is stronger.

Anything on the surface of a large body is in the grip of its gravity, and in order to escape it must move rapidly. If it moves rapidly enough, then even though gravitational pull slows it down continually it can move sufficiently far away from the body so that the gravitational pull, weakened by distance, can never quite slow its motion to zero.

The minimum speed required for this is the "escape velocity." From the surface of the earth, the escape velocity is 7.0 miles per second. From Jupiter, which is larger, the escape velocity is 37.6 miles per second. From the sun, which is still larger, the escape velocity is 383.4 miles per second.

Imagine all the matter of the sun (which is a ball of hot gas 864,000 miles across) compressed tightly together. Imagine it compressed so tightly that its atoms smash and it becomes a ball of atomic nuclei and loose electrons, 30,000 miles across. The sun would then be a "white dwarf." Its surface would be nearer its center, the gravitational pull on that surface would be stronger, and escape velocity would now be 2,100 miles per second.

Compress the sun still more to the point where the electrons melt into the nuclei. There would then be nothing left but tiny neutrons, and they will move together till they touch. The sun would then be only 9 miles across,

and it would be a "neutron star." Escape velocity would be 120,000 miles per second.

Few things material could get away from a neutron star, but light could, of course, since light moves at 186,282 miles per second.

Imagine the sun shrinking past the neutron-star stage, with the neutrons smashing and collapsing. By the time the sun is 3.6 miles across, escape velocity has passed the speed of light, and light can no longer escape. Since nothing can go faster than light, *nothing* can escape.

Into such a shrunken sun anything might fall, but nothing can come out. It would be like an endlessly deep hole in space. Since not even light can come out, it is utterly dark—it is a "black hole."

In 1939, J. Robert Oppenheimer first worked out the nature of black holes in the light of the laws of modern physics, and ever since astronomers have wondered if black holes exist in fact as well as in theory.

How would they form? Stars would collapse under their own enormous gravity were it not for the enormous heat they develop, which keeps them expanded. The heat is formed by the fusion of hydrogen nuclei, however, and when the hydrogen is used up the star collapses.

A star like our sun will eventually collapse fairly quietly to a white dwarf. A more massive star will explode before it collapses, losing some of its mass in the process. If the portion that survives the explosion and collapses is more than 1.4 times the mass of the sun, it will surely collapse into a neutron star. If it is more than 3.2 times the mass of the sun, it must collapse into a black hole.

Since there are indeed massive stars, some of them have collapsed by now and formed black holes. But how can we detect one? Black holes are only a few miles across after all, give off no radiation, and are trillions of miles away.

There's one way out. If matter falls into a black hole, it gives off X-rays in the process. If a black hole is collecting a great deal of matter, enough X-rays may be given off for us to detect them.

Suppose two massive stars are circling each other in close proximity. One explodes and collapses into a black hole. The two objects continue to circle each other, but as the second star approaches explosion it expands. As it expands, some of its matter spirals into the black hole, and there is an intense radiation of X-rays as a result.

In 1965, an X-ray source was discovered in the constellation Cygnus and was named "Cygnus X-1." Eventually, the source was pinpointed to the near neighborhood of a dim star, HD-226868, which is only dim because it is 10,000 light-years away. Actually, it is a huge star, 30 times the mass of our sun.

That star is one of a pair and the two are circling each other once every 5.6 days. The X-rays are coming from the other star, the companion of

HD-226868. That companion is Cygnus X-1. From the motion of HD-226868, it is possible to calculate that Cygnus X-1 is 5 to 8 times the mass of our sun.

A star of that mass should be visible if it is an ordinary star, but no telescope can detect any star on the spot where X-rays are emerging. Cygnus X-1 must be a collapsed star that is too small to see. Since Cygnus X-1 is at least 5 times as massive as our sun, it is too massive to be a white dwarf; too massive, even, to be a neutron star.

It can be nothing other than a black hole; the first to be discovered.

---------------------------------34-------------------------------

# Faster Than Light

In 1905, Albert Einstein worked out his Special Theory of Relativity. One of the basic consequences of the theory is that the speed of light in a vacuum (186,282.4 miles per second) is the absolute limiting velocity we can measure for anything possessing mass — which means any material object we know. That includes ourselves and our spaceships.

Can Einstein's theory be wrong? Not very likely. In the past three-quarters of a century, any number of measurements and any number of investigations have backed it up. The universe acts in the way that Einstein's theory says it acts, and the limiting nature of the speed of light would seem to be as factually solid as the earth we stand on.

But the speed of light is very slow. It seems fast to us on an earthly scale. Anything moving at the speed of 186,282.4 miles per second can move from San Francisco to New York in one-sixtieth of a second and can circumnavigate the globe in one-seventh of a second. At the speed of light an object can go from the earth to the moon in one and one-fourth seconds, and from the earth to the sun in 8 minutes.

But let's get away from the earth and its neighbors. The slowness of light then becomes apparent at once. At the speed of light, any object would take 4.3 *years* to reach Alpha Centauri, the nearest star; 540 years to reach the bright star Rigel; 30,000 years to reach the center of our galaxy; 80,000 years to reach its far edge; 2.3 million years to reach the Andromeda galaxy; and more than 10 billion years to reach the farthest known quasar.

Where does that put science-fiction writers who want to talk of a Galactic Empire, with millions of stars all forming a great community of intelligent beings? Where does it put "Star Trek," with the great starship Enterprise wandering among the stars to uphold justice and put down villainy?

Nowhere! That's where it puts them. We can't have a real social community if it takes thousands of years to travel from one unit to another. Captain Kirk and Mr. Spock would be confined to just a few neighbor-stars for all their lifetime.

What do fiction writers do? If they really know no science, they disregard the speed-of-light limit because they never heard of it.

Better writers know of it and get around it by assuming that in the future, new technologies will be available. They talk about moving through "hyperspace," or through "subspace"; they make use of a "subetheric drive" or a "space warp."

These are just sounds, of course. No one can pretend to describe what such concepts or devices are or how they work. In fiction, though, that does not matter. Such notions at least show that the writer is a respectable craftsman who understands the rules of the universe as set forth by science, and they do make Galactic Empires and starships possible.

But does science hold out hopes that someday there may indeed be detours around the speed-of-light limit?

Yes, but very dimly.

For instance, I said that anything possessing mass has the speed of light as a limiting velocity — but not everything possesses mass. Certain particles, such as "photons," which make up light, X-rays, radio waves, and so on, have what is called "zero rest-mass." Anything with zero rest-mass can move, in a vacuum, *only* at the speed of light, and not the smallest trifle more or less.

Some scientists have speculated that it might be possible for an object to possess the kind of mass that would be represented by what mathematicians call an "imaginary number." If such mass is fitted into Einstein's equations, the results describe objects that can only move *faster* than light. They do not behave as ordinary objects would. The less energy they have, the faster they go until, when they have no energy at all, they move at infinite speed. The more energy they have, the slower they go, until, with infinite energy, they slow down to the speed of light.

Such faster-than-light objects are called "tachyons" from a Greek word for "fast." They are pronounced TAK-ee-onz.

Do tachyons really exist? There is much argument about this, but the only way of ever really proving that they do exist is actually to detect one. This would be very difficult, since any tachyon passing by is likely to be in your vicinity for only a trillionth of a second or less — but it might not be impossible.

So far, however, no tachyons have been detected.

Suppose tachyons *are* detected, though. How might they help us move faster than light?

Well, it is possible to change one subatomic particle into another (obeying the various rules of the universe in doing so), and you can change a particle with mass into one with no mass. For instance, if an electron and positron combine, both disappear. Left in their place are photons. The electron and the positron may have been moving at ordinary speeds, even very slowly perhaps, but once the photons are formed they instantly move off at the speed of light.

Suppose there are ways of changing ordinary particles into tachyonic particles. As ordinary particles, they would be moving at ordinary speeds; but as tachyonic particles, they would be moving faster than light, perhaps millions of times faster than light. Then, if those tachyonic particles were converted back into ordinary particles, they would be moving at ordinary speeds again; but they might be hundreds of light-years away from where they had orginally been, having moved there in a fraction of a second.

Is it possible, then, that someday we might have a "tachyonic drive" that would do all the things that "hyperspace" is suppose to do? Will the Captain Kirks of the future simply shift their ships into tachyonic drive, and will a tachyonic ship then streak swiftly across the galaxies, till it is thrown back into ordinary drive?

It is nice to think about, but there are enormous difficulties in the way.

Even if tachyons exist, no one knows what kind of objects they might be. We might suppose that, for every ordinary particle making up our ordinary universe, there might be a corresponding tachyonic particle in a tachyonic universe. For every proton, electron, and neutron here, there could be a tachy-proton, tachy-electron, and tachy-neutron there. Our particles would make up objects here; tachy-particles might make up tachy-objects there.

But, even so, we haven't the faintest idea of how one would go about changing particles to tachy-particles and back.

And, if we could, we have to remember that we would have to change all the particles at the same time. In order to change the *Enterprise* and *Tachy-Enterprise,* every subatomic particle making up the ship, the cargo, and the crew must change over at the same precise instant. If some changed just a millionth of a second before others did, there would be time, at tachyonic speeds, to spread the spaceship over a distance of billions of miles; and, when all that was changed back, we would have a thin powder of matter, with perhaps some small lumps in it — but we would not have an intact ship and a living crew.

Yet these apparent difficulties may arise simply out of our present ignorance. If we ever detect tachyons and learn enough about them, a tachyonic drive might be worked out very simply according to principles I can't possibly even imagine right now.

Anything else?

I said at the start that every observation in the past three-quarters of a century has backed up Einstein's theory and the speed-of-light limit. These observations, however, are only those it is possible for us to make. There are some observations we can't possibly make. We can't observe the exact conditions at the center of a star, or in a quasar, and we can't make precise measurements at a distance of 12 billion light-years. Can there be any places or conditions where the speed-of-light might not hold?

What about a black hole?

A black hole exists when matter has been compressed so tightly into so small a volume that the gravitational intensity in its immediate vicinity becomes large enough to prevent anything from getting away. That is what makes it a "hole." Even light can't get away, which is what makes it a *black* hole.

Well, then, what are the rules of the universe in a black hole? Are they the same as elsewhere, or are they modified?

How can anyone tell? Astronomers can't study black holes in detail. They aren't even certain they have detected any at all, and those objects that *may* be black holes are thousands of light-years away.

All that scientists can do is try to work out what the rules of the universe *might* be in black holes, by using such basic equations as those of the relativity theory, the quantum theory, and so on.

Some scientists suggest that under some conditions anything falling into a black hole may come out in another part of the universe and it might do so in a very short period of time. In other words, by going through a black hole, it might be possible to go faster than light.

The trouble with that is you would not be able to go wherever you want to. You must go into one end of a black hole, wherever it might happen to be located, and out the other, wherever *it* might happen to be located. It is as though the universe were pictured as consisting of uncounted numbers of cosmic express subway lines, each going from one fixed point to another fixed point, with no necessary convenient relationship between the lines.

Perhaps the *Enterprise* could have a cosmic subway map so that Mr. Spock could figure out which lines to take and which transfer points to use in order to make it from Deneb to Betelgeuse.

But even if that could be worked out, there is this huge difficulty—

Approaching a black hole produces tidal effects so strong that they would reduce any known material to a fine powder, and there is nothing we know of that can afford insulation or protection against such effects. How it would be possible for anyone or anything to make use of the black-hole subway-line without being utterly destroyed in the process is something we can't yet imagine.

So there you are.

Faster than the speed of light? Maybe.

But from where we sit here and now, it's a very, very weak maybe.

# 35

# Hyperspace

In modern science-fiction stories, it is frequently necessary to move quickly across large stretches of the galaxy. Ordinary speeds attained in ordinary ways are insufficient for the purpose. Something special is needed so the spaceship is made to pass through something called "hyperspace" and in the twinkling of an eye — or sometimes in several twinklings — we move away from the sun and are in the neighborhood of some star that is dozens of light-years away.

But what is hyperspace? Is it a term that science-fiction writers have invented out of nothing?

Well, suppose that you were in a vehicle that could move along a single line and could never leave that line. You could imagine a train, for instance, moving along a railroad line with no turn-offs. It could move forward or backward along the line, but in no other way. Another example is an elevator, which can move up or down but no other way.

If the vehicle is at a standstill and you wish to locate it, you need give only one number. If you say that the elevator is 27 meters above ground-level, you know exactly where it is. If you say the train is 175.4 kilometers from the eastern terminus, you know exactly where it is.

Because movement that is restricted to a line makes it possible to locate an object by giving only one number, a line is said to be one-dimensional.

But suppose you were wandering over a large, flat field. You could walk in a straight north-south line, or in a straight east-west line, or anything in between. You can change direction freely, turning right or left to whatever degree you wish. The same is true for a ship traveling over the trackless ocean.

In that case, you cannot locate an object with a single number.

Suppose your home is located in some particular spot and you have wandered off. You reach a telephone, call home, and ask that someone come to get you. You are asked where you are and you answer, "I am exactly 3.58 kilometers from home."

That is not enough. You will be asked in a very exasperated tone of voice, "Yes, but in which direction?"

What is needed now are two figures. You might say, "I am 2.12 kilometers to the north of the house and 2.885 kilometers to the west."

Now someone can drive 2.12 kilometers north from the house, then 2.885 kilometers west, and there you will be. Or else he might cut directly along the diagonal line, a little to the west of northwest. In order to know the direction exactly, he would have to know (or calculate) the precise angle that direction makes with the north-south line or the east-west line.

Speaking on the telephone, you might have said, "I am exactly 3.58 kilometers away from home in a direction lying west of north by an angle of 52.75°." Again you would have to give two numbers, this time a distance and an angle, and that would be sufficient to locate you.

Any part of the surface of the earth, or the whole of it, can be mapped out on a piece of paper. The map is marked off by two sets of lines that meet at right angles—the parallels of latitude and the meridians of longitude. One line in each set is called zero and the others of each set are numbered from zero in an agreed-on way. Once that is done, any place on the earth can be located by means of its latitude and longitude.

Thus, if one were to travel to the spot on the earth indicated by 48.08° north latitude and 11.35° east longitude, one would find oneself in the middle of Munich. Those two numbers would be all the directions needed.

A surface on which one can locate any spot by the use of just two figures is "two-dimensional."

It is not difficult to imagine a third fundamental direction. There is not only north-south and east-west, there is also up-down. We are so used to being bound to the surface that we often ignore up-down in considering location. Suppose, however, we were trying to locate a fly in a room at some particular moment, or an airplane in the air, or a satellite in orbit.

It would then not be enough to give the usual two measurements. You could say, "The plane is over a point on the earth that is exactly 2.55° north latitude and 121.43° west longitude."

The answer would be a testy, "Yes, yes, but how high above sea-level is it?"

We would need a third number.

With that third number we could locate any point in a room, not only from front to back, or side to side, but also from floor to ceiling. We can locate any point on the earth, not only on its surface, but anywhere in the atmosphere, or in the ocean depths, or within the solid ball of the planet itself.

In fact, with three numbers, we can locate any point in space from here to the farthest galaxy, provided we agree on some zero point from which to measure.

Space, therefore, is three-dimensional in this sense.

Can we ever need four numbers? Yes, of course. If we are locating a fly in a room, or an airplane in the atmosphere, or a satellite in orbit, three numbers will do only if we consider a particular moment in time. If you don't do that, then by the time you get the location and look for the located object, the fly or the plane or the satellite is no longer there. It has moved. You need a fourth number to give you the particular time at which the other three numbers are valid.

In that sense, time is a fourth dimension and, in Albert Einstein's view of the universe, time must be taken as an integral part of space; so we speak of four-dimensional space-time.

Time, however, differs in fundamental ways from the other three dimensions.

The dimensions north-south, east-west, and up-down are interchangeable. Suppose you have a cubic box and wish to locate a point within it. You don't have to keep the box in some fixed position. You can move it so that what was north-south becomes east-west and vice versa; or so that what was east-west becomes up-down and vice versa.

For that matter you can make the three dimensions arbitrary; just draw any two sets of lines at right angles to each other, and a third set at right angles to both the first two sets. It doesn't matter at all if all the sets slant in such a way that are none are exactly north-south, east-west, or up-down. Orient the cube in that fashion and the sets of lines will still do to provide three numbers to locate a point.

The dimension of time cannot be handled in this way. No matter how you twist and turn a cube, the east-west line never becomes the yesterday-tomorrow line and vice versa. Nor can the north-south line or the up-down line become the yesterday-tomorrow line.

Then, too, we needn't move in the north-south line, or the east-west line, or the up-down line if we don't wish to. We can stay at rest with respect to them. Or we *can* move, quickly or slowly, as we wish.

In the case of time, on the other hand, we cannot rest; we cannot remain at one point. We travel always away from yesterday and toward tomorrow, all of us and everything, at what seems to be a fixed speed.

Consequently, we can speak of time separately from the other three dimensions. We can say that four-dimensional space-time consists of time and of three "spatial dimensions."

In that case, we can ask whether there is, or can be, a fourth *spatial* dimension? Is there any occasion when we need four numbers to locate an object in space at some fixed time?

No one has ever come across any such occasion, or discovered any case where four numbers are needed to locate a point if time is left out of consideration.

Mathematicians, however, can easily deal with imaginary objects such that the points within them require four numbers for location, or five, or fifty-five, or millions.

Imagine, for instance, a cube that is ten centimeters from left to right, ten centimeters from front to back, and ten centimeters from top to bottom. We know it has six faces, twelve edges bounding those faces, eight vertices where the lines meet. We can calculate the areas of the faces and the volume of the cube.

Now imagine that the cube had another kind of extension; not only left and right, front and back, top and bottom, but also something else we might call hither and yon.

We can't describe this new direction or build a model that shows it, but we can imagine it exists. Within such a cube, with four different kinds of extension, four numbers are required to locate a point at a given time. The imaginary object has four spatial dimensions.

Mathematicians can easily show that such a four-dimensional object is bounded by eight cubes. The total number of faces, edges, and vertices can be counted, and various lengths, areas, and volumes can be calculated.

A four-dimensional object such as the one I have been describing is called a "hypercube," where "hyper" is from a Greek word meaning "beyond," and its total volume is its "hypervolume." You can go up the scale of dimensions: a one-dimensional line, a two-dimensional square, a three-dimensional cube, and a four-dimensional hypercube.

In the same way you can have a one-dimensional arc, a two-dimensional circle, a three-dimensional sphere, and a four-dimensional "hypersphere."

Cubes, spheres, and other three-dimensional figures exist in a three-dimensional space in which three figures (plus a fourth for time) can locate any point. Therefore hypercubes, hyperspheres, and other four-dimensional figures exist in a four-dimensional "hyperspace" in which four figures (plus a fifth for time) can locate any point.

Now that we know what hyperspace is, why do science-fiction writers use it?

Since 1905, unfortunately, those science-fiction writers who know something about science have had to live with a very serious limitation. In that year, Einstein, with his Special Theory of Relativity, showed that the speed of light is the maximum speed any known object or phenomenon can experience.

The speed of light is very great by earthly standards, for it is 299,792.5

kilometers per second; but, considering the size of the universe, it is the merest crawl.

Fast as light travels, it takes it 4.3 years to reach the nearest star, Alpha Centauri; 430 years to reach the bright star, Deneb; 30,000 years to reach the center of this galaxy; 100,000 years to go from one end of the galaxy to the other; 2.3 million years to reach the Andromeda galaxy, the nearest large one outside our own; 1 billion years to reach the nearest quasar; and more than 10 billion years to reach the farthest quasar.

It is possible to write stories in which the long periods of time it takes to go from star to star is an important part of the plot, but, in most cases, science-fiction writers don't want to have their heroes and heroines spend their lifetime in traveling. They'd rather go from star to star in a couple of weeks at most—but relativity won't let them.

But, then, perhaps it is a matter of dimensions.

For instance, suppose we are traveling by ship, back and forth along a river. This is one-dimensional travel; and to go from town A to town B along that river may mean a trip of 100 kilometers.

The river, however, could be a meandering one. Perhaps between town A and town B the river makes a huge curve. If only we could leave the ship at town A and cut cross-country, we could reach town B in a trip of no more than 10 kilometers. In that case, by shifting from one-dimensional travel to two-dimensional travel, we reach our destination much more quickly.

Or else, suppose we travel two-dimensionally along the surface of land or water. Ships must make their way through the viscous water, which is hard to push aside. Land surfaces are rough and uneven. Fifty kilometers an hour is a great speed on water, and two hundred kilometers an hour is a great speed on land. We might feel that moving at still greater speeds is quite impractical.

However, as soon as we switch to three-dimensional travel and move through the air, much greater speeds become possible at once. A supersonic plane can travel 3,000 kilometers an hour. Once again, then, by shifting from two-dimensional travel to three-dimensional travel, we reach our destination much more quickly.

Perhaps the analogy will hold for one more move upward. Perhaps the rules of relativity apply only to three-dimensional space, and in hyperspace *any* speed is possible. At least science-fiction writers can pretend this is so and get on with their stories.

But now that we know what hyperspace is and why science-fiction writers use it, the next question is: Does hyperspace exist in reality?

Unfortunately, as far as we know, it does not.

Mathematicians can imagine hyperspace, and can study, with total confidence, its geometric properties. Science-fiction writers can imagine hyperspace and can, without any worry, invent convenient physical properties. There is no sign, however, that hyperspace exists outside the mathematical or literary imagination. There is no evidence for its real existence, not the slightest—at least so far.

# 36

# Beyond the Universe

It is human to be curious. It is human to be plagued forever with the over-whelming desire to know what lies beyond.

What is on the other side of the hill?

What lies over the ocean?

What exists on the far and hidden side of the moon?

As the centuries have passed, we have moved over every hill on Earth, and crossed over every ocean. We have taken our cameras to the hidden side of the moon and much farther than that, too. We have even taken close-up photographs of distant Saturn.

Our great light-telescopes and radio-telescopes have probed outward for a billion light-years and more, until (we strongly suspect) our instruments have brought us to very nearly the utmost distance we can see.

We believe, in other words, that the universe is finite, that it stretches out only so far. And, if that is so, how can we avoid asking ourselves: What lies beyond the universe?

It would be easy to answer, "We don't know," and end this chapter right now. How unsatisfying that would be, however. Surely we might talk about the matter for a while and do a little thinking about it. If, at the end, we must still say "We don't know," we may at least have the benefit of having placed the problem in perspective a little. We may end by knowing a little more about how and why it is that we don't know.

At any rate, we can try.

One of the problems about trying to decide what is outside the universe is that the universe is so large. The most distant objects we have detected are over ten billion light-years away from us, and the end of the universe—together with what lies beyond it—must be farther still, and therefore extraordinarily hard to reach and study.

What if the universe were smaller, however—even much smaller? Might it not have been easier, then, to detect its boundary and look beyond it?

Actually, there was a time when the universe was smaller, even much smaller, than it is now.

Right now, the various clusters of galaxies are moving away from each other so that the universe as a whole is expanding. It has been expanding for thousands of millions of years, and it will continue to do so for thousands of millions of additional years.

If the universe was and is and will be continually expanding, then it is larger now than it was yesterday; and it was larger yesterday than it was day before yesterday, and so on.

If we imagine ourselves traveling backward in time, we would see the universe steadily contracting and contracting. If we move far enough back in time, the universe would contract to quite a small volume, perhaps even to a very small volume—to a virtual pinhead.

The evidence seems to show that this is so; that thousands of millions of years ago all the matter and energy of the universe was concentrated into a small body and that this small body exploded with an unimaginably great burst of energy at incredibly high temperatures in something called the "big bang." The temperature of the exploding universe rapidly decreased after the big bang so that the sea of energy that first existed coalesced into matter. The matter collected into galaxies, which in turn coalesced into billions of stars in each. Eventually, the universe as we now know it evolved, continuing to cool off steadily over the billions of years. Today, it is very large, quite cold, and still expanding.

Need we feel bitter regret that we didn't live billions of years ago at a time when we might easily have used our instruments to penetrate beyond the nearby edge of a pygmy universe and see what lay beyond?

No, for it would have been just as difficult to do that then, however small the universe, as it is now. To see why that is, let's consider a simple analogy to the case of the expanding universe.

Imagine a balloon being blown into and made to grow slowly larger. We can suppose that it started very small indeed and that it could grow larger and larger without any danger of breaking. We can further suppose that a small region of the skin of the balloon is a home for submicroscopic intelligent creatures that are forever confined to that skin.

Let us imagine that the creatures of the skin of the balloon can travel freely anywhere through the skin but can never venture away from it either outward or inward. What is more, the only effects they can detect either by their senses or by their instruments are confined to that skin. The rays of light they observe, for instance, follow the curve of the skin and never leave it.

The entire universe of the creatures is the skin of that balloon.

If the balloon is large, and if the creatures are submicroscopic in size and can move only at submicroscopic speeds, they can only directly observe a very tiny patch of that balloon-skin, and they can be forgiven for thinking the skin is flat. Over a tiny patch, it *is* just about flat.

Imagine tiny imperfections of some sort scattered more or less evenly through the skin of the balloon. As the creatures grow more and more sophisticated, they would build instruments that could detect these imperfections at greater and greater distances (enormous distances compared with the portion of the skin they can study with their unaided senses). As they do this, they might well discover certain subtle effects that would convince them that the skin of the balloon is curved.

Furthermore, as they study the distant imperfections, they begin to notice that each imperfection is getting farther and farther away from themselves. The more distant an imperfection is, the faster it is receding. In fact, all the imperfections are receding from each other.

The conclusion that the creatures would come to is that their universe is expanding.

Mind you, they don't necessarily conclude that they are part of a balloon that is being blown up and made larger. They know nothing about the balloon as a whole. All they are aware of is the skin of the balloon. But, then, as you blow up a balloon, the skin must stretch to surround a larger and larger volume of space within. The skin itself expands therefore as the balloon gets larger, and it is this expansion of the skin that the creatures observe.

Suppose that the circumference of the balloon, at a particular instant of time, is two meters. The creatures will notice that all the imperfections they observe are never more than one meter away from themselves. Their universe is therefore finite in size and the creatures may begin to wonder: If the universe is finite in size, what lies beyond it?

By the terms of our suppositions, this is a question to which they can never get an answer. If they develop instruments that can detect imperfections that are exactly a meter away, those imperfections are on the opposite side of the balloon, on the point precisely opposite to themselves. Anything that is further away than that from the observation point, when viewed in one particular direction, is actually closer than that when viewed from another direction. If something is detected that seems two meters away in a particular direction, the light it emits has had to travel around the full circumference of the balloon. If the observer about-faced and looked in the opposite direction, he would find himself close enough to the object to touch it.

Again, if the creatures invented some sort of vehicle that carried them through the skin of the balloon at what to them would be enormous velocities, it would take them farther and farther from home, until they were one meter away and at a point on the balloon on the precise opposite side—the antipodal point. If from that point they continued onward in what seemed to them to be the same straight line (or, for that matter, in *any* direction), they would find themselves getting closer to home. A trip of two meters, in what would seem to them to be a straight line, would bring them home again.

You see that, in this way, they could keep on traveling forever, round and round the skin of the balloon, without ever coming to an end, even though they were correct in thinking the skin was of a finite size. —And they would never discover what lay beyond the edge of a universe whose edge they could never reach or find.

And this would be true not only when the balloon was expanded to a large size, but even when it had been very small. To those confined to the skin, there would be no end, however small and unextended the spherical skin might be.

The surface of the earth itself offers a similar case.

Imagine that human beings are absolutely confined to the surface of the earth; that they can travel freely in any direction upon its surface, but could never rise upward or dig downward; that they could never receive any information in any way about what lies overhead or underfoot.

If their experience was then confined to only a comparatively small area of the earth, it would seem flat to them and they would imagine that somewhere there would have to be an end. They might then legitimately wonder what lay beyond that end. (As a matter of fact, in early times, that was exactly the thought human beings had.)

An understanding of the true nature of the earth's surface, based on purely surface observations, began to come when it was noticed that ships disappeared hull-first as they approached the horizon. Full appreciation of the earth's spherical nature did not come till long ocean voyages became common, after 1400, and it turned out that ships could not be guided correctly unless the curvature of the earth was allowed for.

Then, too, if one measured along the surface of the earth, a human being could never find himself more than 20,000 kilometers from home. If he then tried to go farther away still, in *any* direction, he would find he was *approaching* home.

The earth's surface was finite all right, but there was no end to it. A traveler could go on forever, this way and that at will, and find no end. If he confined himself to the earth's surface, the question "What lies beyond the end of the earth?" has no answer because there is no end, even though the earth's surface is finite.

Now let's consider the universe. The universe resembles the skin of the balloon and the surface of the earth, except that the universe is three-dimensional, while the skin and the surface are essentially two-dimensional. The skin of the balloon and the surface of the earth are two-dimensional objects curved through a third dimension. The universe is a three-dimensional object curved through a fourth dimension.

We are confined to this three-dimensional "universe-skin." We can travel up and down, right and left, forward and backward, or in any combination of these motions, but we can't ever travel out of the universe by way of the fourth dimension.

This means that we can't ever get to that point from which the universe originated in its big bang. That is the center of the four-dimensional balloon, and we are in the three-dimensional ever-expanding skin. The point of origin is equally distant — and equally unattainable — from every portion of the universe (just as the center of the balloon would be equally distant from every portion of its two-dimensional skin and would be equally unattainable from every portion of it as far as any creature was concerned who was absolutely confined to that skin).

In the same way, if we could imagine ourselves traveling through the universe at a speed of light-years per second, we could reach the farthest known object in it in a year or so, perhaps, and then find we were no nearer the end of the universe. We could get only so far away from home and then, in whatever direction we traveled from that most-distant point, we would get closer to home. Just as in the case of the skin of the balloon or of the surface of the earth, we could wander on forever in any direction we chose and would come to no end, even though the universe, like the skin of the balloon or the surface of the earth, is finite.

Of course we can't move light-years per second. The highest conceivable velocity (by current scientific views) is that of a ray of light traveling through a vacuum. That moves at only 300,000 kilometers per second, takes a year to travel a distance of a light-year, and would take thousands of millions of years to reach us from some very distant object.

If we look at a very distant quasar, the light by which we see it left it some ten billion years ago. We are therefore seeing it as it was ten billion years in the past. At that time, it was part of a universe that was much smaller than the one we live in now. If we imagine ourselves traveling backward along the path taken by the ray of light, we would find ourselves following a four-dimensional route backward through time, one that spiraled inward as the universe contracted.

We can't do that, however. We can only travel at the speed of light or less, and if we moved in the direction of that distant object the universe would expand as we traveled and we would follow an outward spiral. (It would be a four-dimensional route forward in time, but one that carried us forward willy-nilly in a movement through time dictated by the expanding universe, which we cannot alter or modify in any way.) When we finally reached that distant object many thousands of millions of years from now, the universe would be much larger than it is now.

Larger or smaller, however, whether we spiraled forward in time or backward in time, we would nevertheless remain in the universe wherever we went, and at no time, either in the far past or the far future, would we come to an end.

No matter how small the universe became as we imagined ourselves moving backward in time, there would be no end. Even if we imagined ourselves traveling so far into the past that we would reach the original tiny object that exploded, that tiny object would be the entire universe and we would be penned up within it, trapped in a tiny skin that we could not leave and that had no end.

So you see, it wouldn't matter if we lived in a small universe as far as knowing what lay beyond its edge. We might as well live in the large one we do.

But wait. The creatures of the skin of the balloon can find no end only as long as they are confined to the skin. If they could move outward or inward from the skin, they would pass beyond the edge of their universe and find that beyond that edge is air.

Again, human beings confined to the surface of the earth can find no end; but, if they move in the third dimension, they move beyond the surface-universe and find that beyond in the upward direction is air, and then, eventually, the vacuum of space. Beyond in the downward direction is rock and then, eventually, molten metal.

We might say then: "Never mind the confines of the three-dimensional universe. What exists beyond the boundaries of the universe in the direction of the fourth dimension? What lies in those four-dimensional regions into which the universe has not yet expanded? What lies in those four-dimensional regions out of which the universe has long since expanded?"

Since, as far as we know, nothing reaches us from those extra-universal regions, we have no evidence concerning them. There is, therefore, nothing we can know. We can only guess.

Suppose there is nothing outside the universe. Really *nothing*. I don't mean a vacuum, but truly NOTHING.

We are always talking of the vacuum of outer space, the emptiness between the stars and galaxies. What we call a vacuum, however, is far from nothing. It may not be as full of material particles as our immediate surroundings are, but even those places farthest away from any stars — deep, deep in the "emptiness" of space — probably contain at least one material subatomic particle per cubic meter.

Besides, every particle of matter produces a gravitational field, an electromagnetic field, a strong nuclear field, a weak nuclear field — or some combination of them. Of these, the gravitational and electromagnetic fields are long-range and can exist in measurable intensities even at astronomic distances.

Every smallest portion of space, then, however far from any material object it may be, is constantly criss-crossed by gravitational waves and electromagnetic waves. They are also criss-crossed by certain immaterial particles called neutrinos. If we include these waves and immaterial particles, then the universe is full and will always be full no matter how large it gets and no matter how thinly the material portion of it is spread out.

If we pass beyond the universe in the direction of the fourth dimension, however, suppose we enter a region in which there is not only no matter but no immaterial particles either, and no fields and no waves—only NOTHING.

How could we possibly study the properties of nature of such NOTHING-ness? The instant we, or our instruments, emerged into it, we or they would serve as the foci for gravitational and electromagnetic fields spreading outward in all directions at the speed of light.

In other words, any attempt we made to study NOTHING would convert it at once into ordinary space. Even if we left the universe, we would carry the universe with us and manufacture new universe about us and *still* never reach the end.

But suppose beyond the universe is not NOTHING, but SOMETHING, and that the universe in mixing with the SOMETHING would change its nature and give us something new to study, something that would tell us what lies beyond the universe.

However, the universe, in that case, as it expands is always mixing with SOMETHING and ends, in the process, as the universe we study. If we go beyond it, we will extend the mixing and, again, produce merely more universe.

Whatever we do, then, we will carry the universe with us; we can never leave it; there is no end even though it is finite, and the question as to what lies beyond that end is meaningless.

---37---

# Life on Earth

The universe is inconceivably large. It may be made up of as many as one hundred billion galaxies.

Each galaxy contains anywhere from a few million to a few trillion stars. Our own galaxy, the Milky Way, has about three hundred billion stars glittering within its core and its spiral arms.

Of all this, our sun is but one star; one star lost in vast crowds.

Surrounding the sun is a family of smaller bodies, billions of them, ranging in size from the largest planet to the smallest asteroid or comet. (Perhaps every star has such a family.)

Our earth is just one of all those objects circling the sun and is not even the most impressive. It is the fifth largest.

It may be that only on our earth, on this one world in an enormous universe, is there to be found the phenomenon of life.

There may be other worlds full of life, to be sure, innumerable examples of them dotting the universe—but we don't know of them. The few nearby worlds we have reached and examined seem lifeless.

Is there any point in studying life? If it is such a small part of the scheme of things, can it have any importance? After all, if all of life on the earth were to vanish suddenly, the planet would continue to spin and circle the sun and would do so unperturbed for billions of years. The sun certainly would continue shining for those billions of years, in no way affected by the disappearance of living things on one of the many objects circling it.

And the rest of the universe? It could not be altered or touched in any way by that disappearance.

Still—we are part of life, so we may be excused if we find selfish cause to be interested in it and to find it a wonderful phenomenon.

But leave self-interest out of it! Imagine yourself a disembodied intelligence having no connection with life whatsoever. In that case, why study *it* rather than the overall structure of the universe? Why be concerned with some worm or fern or kernel, instead of finding fascination in a star, a supernova explosion, a black hole?

Easy. Life, as far as we know, is the most complex phenomenon that exists. Nowhere, in anything nonliving, are atoms arranged into such complex molecules, molecules arranged into such complex organisms, organisms arranged into such complex social systems.

An amoeba is more complex than a star and less predictable. An astronomer can, with reasonable confidence, predict just what a star will be doing one billion years from now, while a biologist can only guess what an amoeba will be doing fifteen minutes from now.

This means that the phenomena of life are more interesting to observe, more fascinating to study, more challenging to discover, and more rewarding to understand than anything else can be.

This would be so if we were studying only a few varieties of life, but we are not. We are studying a planet that is full of life.

There are a million species known, and thousands of new species are reported every year. There are probably two million species living on the earth right now, altogether—each one distinct from every other—each one with its properties, abilities, adaptations, techniques of survival—each one playing a special part in the overall scheme; eating or being eaten, pursuing or fleeing, enduring or dying.

Life would be fascinating, even if it were restricted to some particular environmental niche, but it is found everywhere: in hot springs and in arctic pools; in deserts and in the oceanic abyss; in bogs, and plains, on slopes, along tidal estuaries. And whatever the nature of the environment, the living things who exploit it show, in every part of their form and function, a design that serves to make that exploitation efficient.

Life would be absorbing even if it were strictly utilitarian in all its aspects, but it is more than that. There is surprising beauty even where one would suspect beauty to be superfluous. There is symmetry, grace, color, and artistry in things as different as the petal of a rose and the feather of a bird. There is scarcely a living motion that does not seem choreographed, a living sound that does not seem orchestrated, a living form that does not seem painted or chiseled.

Life would be astonishing even if it included only those forms we could see, but the microscope has opened up a whole new world of life, showing us tiny creatures, each as alive as we are, and each as finely designed for survival as we are. There are creatures so small that they consist of scarcely more than a few large molecules, and yet they compete for survival successfully with those creatures, also alive, who are as large to them as the earth is to us.

Life would be impressive even if interrelationships existed only in the present, but they do not. They stretch backward in time for billions of years and, behind the mighty list of species that now exist, there are perhaps ten times as many that once existed and that exist no longer. And among those species, now forever gone, were organisms larger and physically more impressive than almost any that live now. The interrelationships across time are at least as interesting and as significant as those across space.

Life would be stupendous if its tens of millions of species, present and past, pursued their life-cycles physically unrelated to each other, but they do not. The species are connected, each one to all the others, not only in behavior and interdependence, but physically. Life, with all its unimaginable variety, is a unit. What seems like a vast sea of separate lives and concerns when viewed at this instant of time, becomes an even vaster "tree" when viewed through the history of our planet.

From its earliest beginnings, the initial forms of life have varied—mutated—lived and flourished or dwindled and died. The whole panorama of change and of selective survival is the drama of biological evolution, which has produced the human being—and the tapeworm—each equally adept at survival and each equally a biological success story.

Life would be awesome even if human beings could step aside and view it without personal concern, but they cannot. *Homo sapiens* is one species of millions, and every one of the others is our concern. There are species of plants and animals that serve as food for us. There are life-forms that fertilize our soil and pollinate our plants; others that scavenge the dead and recycle the refuse; still others (that lived long ago) that have supplied us with trillions of tons of coal and billions of barrels of oil. All of the species form an intricate lacework, whose interrelationships we have not yet fully plumbed, but on which, we know, we must depend.

Life would be colossal even if all I have said about the present and the past were all there was to be concerned with; but all that is happening and all that has happened is sure to be dwarfed by what is to come.

Given the worst scenario, it may be that human beings will continue to use their restless intelligence to pursue their busy self-interest without thought of consequences. We will multiply our population and level the wilderness, and deplete the soil, and destroy habitats, and drive other species to extinction by the dozens, then by the hundreds, then by the thousands.

We will deplete the ecological pattern, damage the intricate network of interrelationships, gobble up food and resources for endlessly increasing mouths and hands, make life ever harder for our own increasingly isolated species, and finally destroy our civilization, ourselves, and much of life in the hell of a nuclear holocaust.

If this is not to happen, it will be, in part at least, because we will study life soberly, not merely our own life and health, as though humanity itself were all that counted, but the life of the planet as a unit.

There is a great deal to learn in addition to the mere description of appearances and habits and adaptations and ways of life. We must burrow within (and we are doing so) to study the interplay of atoms and ions and molecules that blend to form systems so complex and versatile that we can think of them as living.

That it is taking us so long to do so is not surprising. Human knowledge has moved from the simple to the complex.

Modern physics, the simplest of the sciences, received its unifying principle and its modern shape with Isaac Newton's great book, *Principia Mathematica,* published in 1687.

Modern chemistry, a step up in complexity, did not receive its unifying principle and its modern shape until Antoine L. Lavoisier's work on combustion and on conservation of mass, nine decades later, in 1778.

Modern biology, another step up in complexity, did not receive its unifying principle and its modern shape until Charles R. Darwin's book *The Origin of Species* was published, eight decades later still, in 1859.

All three great branches of science have advanced since their beginnings, and biology, dealing with the phenomena of life, has, in its incredible complexity, drawn endlessly on the other two. Physics and chemistry have blended with life science to produce "molecular biology."

This study of the molecules characteristic of life penetrates to the behavior of the enzymes that control all the chemical reactions in the cells, guiding and intermeshing them with a gentle certainty that reduces the most enormous product of merely human ingenuity to the level of the wheel and lever in comparison.

That same molecular biology has also penetrated to the core of the nucleic acids that control the formation of enzymes and that make certain that out of all the incredibly numerous enzymes that could potentially exist, only those characteristic of the particular cell and organism will actually be formed.

It penetrates the manner in which nucleic acids duplicate their own structure to produce other molecules exactly like themselves, so that the endless construction of cells and organisms just like those doing the constructing continues and so that dogs have puppies and cats have kittens, and never vice versa.

It goes further; it penetrates the manner in which nucleic acids, in their remorseless duplication, make a myriad of tiny errors; not enough to destroy the blueprint, but sufficient to introduce all those variations that make it certain that of all the billions of people on the earth, no two are exactly alike, and that you can recognize your friend from among all the rest without trouble. It is sufficient to produce all the variations in all creatures upon which natural selection can seize, in order to promote some and suppress others and to allow evolution to continue its course of increasing variety and of improving the fit of organism to environment.

As yet we are only at the beginning of this sort of knowledge, but we have learned enough to make it possible to see the shape of "genetic engineering" looming up before us in the future.

We have tamed large organisms in the past, but now we may learn enough to tame the giant molecules of living tissue—nucleic acids and enzymes—and to bend them to our service (with wisdom, as well as knowledge, we can hope).

We may learn to reshape the tiny chemical factories of bacteria, molds, and algae, in order to have them do what we will find useful, to produce hormones and other biological substances that can contribute to human health, to scavenge more efficiently those products of human ingenuity that the present ecological balance can't handle properly, to promote the production of fuel and fertilizers, and so on.

We may learn to manipulate the genes of living organisms to increase the versatility, complexity, and strength of the ecological pattern that, so far, we have done so much to destroy. We may even learn to re-design ourselves to remove physical defects and introduce new strengths.

Since Newton's time, we have lived in the Age of the Physical Sciences, but we may now be entering the Age of the Biological Sciences.

In fact, our study of life may now be transcending the earth itself.

We have finally devised instruments capable of detecting occasional bursts of energy, or of speeding particles, from distant sources. It is possible that we may, in this way, gain enough information to tell us whether, somewhere in space, there is a civilization sufficiently advanced and sufficiently near to us to be detectable by the energy it broadcasts (either as a deliberate signal or as an exuberant overflow).

The mere detection of even a single undoubted sign of intelligent life elsewhere would be of enormous importance, if only because it would be an indication that some life-form somewhere has managed to advance technologically beyond the point we have reached, and to do so without destroying itself. It would offer us the heartening possibility that we might be able to do the same.

But, even if no search of ours will net us any signals, if we can find no sign that we are not alone, it is only *intelligent* life that we will decide may be absent. There might still be life in overflowing quantity. After all, on the earth itself, for over three billion years, twenty million species developed (and most of them died out) before a single species developed sufficient intelligence to build a technological civilization.

Then, too, we ourselves will continue (perhaps). Even if there is no life in the universe except for ourselves, we could be enough to correct that fact, as we restlessly and indomitably move outward from the earth toward the distant stars.

It is for this reason—to see and here all that is and has been, and to think of what will be and may be—that it is useful, and delightful, to watch David Attenborough's television series, "Life on Earth." It is an overview made more powerful than anything preceding, thanks to the full application of human technology to this demonstration of the beauty, the variety, the versatility, and the complexity of life.

# Part VI

# The Future

---

# 38

# Transportation and the Future

Motion is as old as life and indeed, is inseparable from life. If an organism doesn't move as a whole, there is still motion of its parts or within its cells.

Motion requires energy, but for almost all species in the history of life that energy has been supplied by the chemical reactions within the cell and, in most animals, by muscular contraction specifically. Human beings, through most of their history, have moved by such internal energy, which makes it possible for them to walk, run, crawl, hop, and jump.

Transportation (from Latin words meaning "to carry across") refers especially to the carrying of objects from one place to another—and the organism itself might be the object carried. This, too, is common in life. Parasites are carried about by the organisms they infest; spores and seeds are blown about by the wind; many life-forms are carried about by water currents; remoras deliberately attach themselves to sharks.

Human beings, like many other animals, carry or drag other objects when it suits them to do so. Before the dim dawn of history, human beings already arranged to have themselves carried, using donkeys for the purpose.

About 2000 b.c., the horse was tamed. First, horses drew chariots, but eventually human beings learned to ride them. For nearly four thousand years, the fastest way in which a human being could travel on dry land was to bestride a galloping horse. By using relays of horses, human beings could cross continents in surprisingly short times, and in the thirteenth century the Mongol armies perfected the art of the blitzkrieg with nothing more than shaggy ponies at their disposal.

On the rivers, the water current took the place of horses. Human muscles at work on poles or at oars forced boats against the current or

191

across the placid sea. When the wind blew in the right directions, sails could take advantage of the energy of moving air.

Until the dawning of the nineteenth century, however, all human transportation was made possible by the energy of (*a*) human muscle, (*b*) animal muscle, or (*c*) the inanimate and uncontrolled motion of water or air.

In 1807, for the first time inanimate energy under the complete control of human beings was applied to the problem of transportation, when Robert Fulton devised and ran the first commercially successful steamship. The energy of burning fuel, which produced the power of expanding steam, was also applied to land travel by George Stephenson, who devised the first commercially successful locomotive in 1825. Thus was initiated the modern era of transportation, and the world began to shrink.

The internal-combustion engine was made commercially practical in 1877 by Nikolaus August Otto. It proved to be much more convenient than the steam engine, and in the century since it has powered automobiles, buses, trucks, trains, ships, dirigibles, and airplanes.

Thanks to the internal-combustion engine, the United States, in particular, has become a nation on wheels, its people forever on the move, its population shifting and homogenizing.

During World War II, the jet to a large extent replaced the internal-combustion engine of the older variety in airplanes, and a version of the jet principle, the rocket, has carried transportation beyond the atmosphere. We have now reached the point where no place on Earth is more than forty-five minutes from any other by rocket, or more than six hours from any other by supersonic jet.

Where is there left to go?

In one way, we might rest content. There is scarcely any practical use, as far as travel on Earth is concerned, in going faster than a rocket in orbit; and it is hard to see how anything can be more convenient for individual use than our beloved automobiles.

Yet we *can't* rest content.

For one thing, there's a price. The vast energies we use for transportation have filled and dirtied the atmosphere with fumes and pollution; have created traffic jams, traffic accidents, traffic fatalities; have filled the world with noise; have produced urban and suburban sprawls.

For another thing (and more fundamentally) the vast energies used came into their own with the widespread use of petroleum (oil), from which is distilled the gasoline and diesel fuel we use by the millions of barrels daily. Our worldwide transportation network runs on oil—and the oil is running out. We are in the same position the eighteenth century would have been in if their horses were dying.

Let's take up these problems bit by bit.

As far as the noise and pollution is concerned, that is closely allied to the internal-combustion engine, which, however, is not the only way of propelling a car. During the automobile's early history, electricity was used to propel some makes of cars. Electrical cars were by far quieter and cleaner than gasoline cars, but the latter won out. Partly, it was because gasoline came to be so cheap, partly because it allowed rapid acceleration and high speeds, and partly because the gas in the tank lasted a long time and was simple to refill, while the battery quickly ran down and was tedious to recharge.

Today, however, gasoline climbs in price, declines in availability, and adds to unbearable noise and pollution — while storage batteries are at last being improved. A nickel-zinc battery, which may become commercially available within two or three years, will store three times the electrical energy, pound for pound, that the time-honored lead battery will.

Since energy shortages are, in any case, enforcing smaller, lighter cars moving at moderate speeds, electric cars will become competitive again. They will be able to reach speeds up to the legal limit of 55 miles an hour and will travel a hundred miles between re-chargings. In ordinary commuter travel, cars can be recharged routinely overnight in the garage. In extended travel, the driver will stop periodically at a service station to recharge or, better, simply to replace the spent battery with an already charged one. (That would be rather like using relays of horses.)

The electrification of cars will reduce the noise level on city streets and highways to below the level of even the horse era with its clanging of iron-shod hooves. The pollution level, too, would be reduced to below the horse era, for there will be neither exhaust fumes nor horse-manure.

What about traffic jams and traffic fatalities? We might fairly suppose that without the unnecessarily high speeds to which the gas-guzzling internal-combustion engine tempts us, there will be fewer accidents. But we need something more.

Here, we will be able to rely on increasing automation and on the versatility of the microcomputer.

The traffic lights of today are inflexible, shifting at prearranged times without any regard for the condition of the traffic. It should become possible in future years to have traffic lights that are capable of scanning the approaches to an intersection so that they can detect the relative density of traffic in the different directions and adjust the relative lengths of time that stop-and-go signals are given. In this way, maximum traffic flow would be made possible. In fact, whole networks of traffic lights might be computerized to react to traffic flow and adjust themselves cooperatively to keep cars moving in the most efficient possible manner.

Individual cars could use radar guidance that would make them capable of detecting obstructions at night, or in fog, and to detect deceleration in an automobile ahead or a car's swerving to one side. It is likely that cars would be equipped with automatic devices that would slow, turn, or halt them in accordance with radar messages.

The automobile of the future may, in fact, be altogether automated. Its computerized "brain" could retain the highway network with which it would be programmed so that the automobile would follow the most efficient route, while guarding against obstructions and other cars and while adjusting its speed to that of the traffic flow. It would require manual operation (or perhaps emergency reprogramming) only in case of unexpected detours.

The more automated a car, the more desirable that it be electrified. An electric current can be started or stopped or shifted from one pathway to another more easily and quickly than the same can be done with a current of inflammable vapor.

One might argue that the electric car merely displaces the problem rather than solving it. Instead of each car's burning fuel on the spot to produce its small share of pollution, electricity will be generated at huge oil- or coal-burning plants, which will produce all the pollution, and more, than the individual cars would have produced.

Not necessarily.

In the first place, it is not unreasonable to suppose that methods for cleaning the fuel will be developed, together with methods for cutting down the emission of undesirable smoke components. It is easier to control and oversee a relatively few electricity-generating plants than hundreds of millions of individual automobiles.

In the second place, there are methods of producing electricity that do not involve the burning of fossil fuels with their production of nearly inevitable air pollution. Hydroelectric turbines produce electricity, and so do nuclear-fission plants. The production of electricity by windmills or geothermal energy is in the cards. In the future, nuclear-fusion plants (much safer, scientists hope, than nuclear-fission plants and far more energy-rich) will do the same, and there will be increasing use of photo-voltaic cells that will turn sunlight into electricity.

The electrification of the automobile is therefore likely to represent a general and absolute decline in pollution.

To be sure, probably not all forms of transportation will be easily electrified. Electric trains are a commonplace, but buses, trucks, ships, and airplanes would represent difficulties.

Even if we grant that the internal-combustion engine will live on in the larger vehicles, however, the use of the electric automobile (assuming that the electricity is ultimately produced by some method other than the burning of oil) will represent an enormous oil-conservation measure. Without automobiles draining the oil supplies of the world, those oil supplies will endure perhaps half a century longer than they otherwise would, and will give us that much more time to develop the alternative energy-sources we desperately need.

There are, of course, alternatives besides the internal-combustion engine and electrification for the powering of vehicles. We already have nuclear submarines, and nuclear surface vessels have been built. There has been some talk of nuclear-powered airplanes and, in the wildly optimistic early days of nuclear fission, there were speculations about nuclear automobiles.

The public perception of the dangers of nuclear fission (which are currently much exaggerated) makes it seem unlikely that nuclear transportation will become a major force. Yet what about fusion vehicles or solar vehicles?

We don't have to imagine the direct application of these novel power sources, however. We needn't imagine airplanes bearing solar cells, or buses with, somewhere in their vitals, heavy hydrogen nuclei fusing under the impact of laser beams. Instead we can consider transportation depending on fusion or sunlight indirectly.

Consider that the plant world can (at the expense of the energy of sunlight) combine carbon dioxide and water to form foodstuffs and oxygen. Some of the foodstuffs are not too far from gasoline in their chemical nature. In fact, there is considerable talk of harvesting some plants for the fuel-worthy alcohols and hydrocarbons that could be obtained from them. Well, given sufficient energy (and either fusion or the sun can supply that) human beings can themselves combine carbon dioxide and water to form liquid fuel and oxygen in any desired quantity.

This will certainly be more convenient than pumping oil out of wells, or baking it out of oil shale or tar sands, or wringing it out of coal. In addition, the liquid fuel formed out of carbon dioxide and water will contain *only* carbon and hydrogen atoms (and perhaps an occasional oxygen atom). It will contain none of the nitrogen, sulfur, and other atoms inevitably found in oil as it is recovered from natural sources; none of the atoms that give rise to the obnoxious pollutants in smog and in acid rain.

When the synthetic fuel obtained from carbon dioxide and water is burned, it will combine with oxygen to form carbon dioxide and water again, and nothing else.

The whole process moves in a cycle. Carbon dioxide and water form liquid fuel and oxygen, which combine to form carbon dioxide and water again. All that is used up is energy from the sun or from fusing nuclei, either of which will last for billions of years, so that for those applications that require liquid fuel there is no danger of ever running out.

So far, the vision I present solves the problems of noise, pollution, traffic jams and accidents, and the drain on our vanishing supply of oil. It ends by leaving the general picture of transportation much the same as it is now — but of course there will be changes.

There might, for instance, be improved methods of mass transportation on the ground, thanks to the use of magnetic levitation.

A central magnetized guide-rail will repel a train that is similarly magnetized. If the magnetic-field intensities are great enough, the train will lift a tiny distance above the rail. The train could move forward under the impulse of an electromagnetic field and, since there would be virtually no friction without solid contact, vibration-free speeds of three hundred miles an hour could be obtained (with appropriate safeguards against jumping the track).

The efficiency of mass transportation might be further increased by moving it underground. We have had subway trains for nearly a century, but it is conceivable that this trend may be accentuated and that whole cities may move underground.

That may seem strange to us, used as we are to open-air existence, but there are advantages to underground living.

The chief of these would be the final defeat of weather, which is primarily a phenomenon of the atmosphere. Rain, snow, sleet, fog would not trouble the underground world. Even temperature variations would not exist underground. Day or night, summer or winter, temperatures in the underground world would remain equable. The only natural danger that would remain to be feared would be the earthquake.

The defeat of weather would be of prime importance to transportation, since that would make popular the basic human device of walking. Without cold and heat, wind and wet, walking short distances would be far more pleasant. For longer distances, there could be moving walks on the level, and escalators and elevators to progress upward or downward. A city would be very much, in this respect, like a gigantic office-building.

Nor would it particularly mean that people would be separated from nature. Rather the reverse.

In a huge metropolis on Earth's surface, people must travel many miles, sometimes, before they find themselves in the untrammeled countryside; in an underground city, a short elevator-ride upward from any point could do.

Subways might then be extended to stretch from city to city. Long-distance travel would be through long tunnels by magnetic levitation. There would be the added refinement, more practical underground than on the surface, that the tunnels would be evacuated. With the absence of air-resistance, trains could move very fast indeed, and transcontinental trips might be made at supersonic speeds as fast as or faster than airplanes could do it.

An expanding network of such tunnels could finally connect all major population centers in the Americas, while a second network could connect the population centers in Eurasia and Africa. Add a tunnel under the Bering

Strait, and there would be interconnections from Patagonia to Portugal and Capetown.

Such a worldwide underground transportation system would have the advantage of being adjusted to a worldwide time-pattern. After all, there would be no natural day-and-night alternation traveling inexorably about the world as the earth rotated. A single time could exist for the entire planet and "jet lag" would be a thing of the past.

To relieve transportation pressure on the earth's surface, it would also be possible to move in the other direction — to lift vehicles *above* the surface of the earth by a small distance. This can be done, if automobiles travel on compressed air jets. As in the case of magnetic levitation, friction would be reduced sharply, allowing greater speeds and less vibration and noise.

Air-jet transport is more flexible in one respect: Surface transport is confined to roads and highways, subsurface trains to rails and tunnels. An air-jet vehicle could, however, move wherever the surface beneath was reasonably flat. Traffic density would, in theory, drop everywhere except at a relatively few bottlenecks, most of which could be engineered away.

There would be disadvantages, however. The compressed-air jets could raise clouds of dust or do damage if the vehicle is not confined to a paved road, to say nothing of the consequences of intrusion upon private property. The coming of such air-jet vehicles, therefore, would require considerable thought as to the need of appropriate regulation.

A particular advantage of the air-jet vehicle is that it could travel over water as easily as over land, in some ways more easily. A water surface is essentially flat everywhere, except in storms, and is not parceled out into privately owned bits.

Since rivers could be crossed at any point, the traffic load on bridges would be decreased.

Oceanic mass-transport might be jet-assisted. Passenger liners and large freighters may use jets to lift them higher in the water, reducing friction and making higher speeds possible. Vessels of small to intermediate size could be lifted entirely free of water (this is already done).

Such air-jet ships might be a major way of linking Australia, Antarctica, and various islands with the subsurface train-system that would honeycomb the major continents.

Such vehicles, equally at home over water and land, would not have to bring their cargoes to the few specialized ports that now exist, but could move directly to whatever portions of the earth's land surface offered them the easiest and most convenient way of delivering their goods.

The airplane, too, could show important advances. Air flight is now a specialized activity that requires airports and runways. Even small planes require an extended stretch of flat surface to take off and land. Helicopters, which are less demanding in this respect, are slower than ordinary planes.

We might, in the future, have Vertical Take-off and Landing (VTOL) planes. Such planes would not require long runways, but could take off from one backyard and land in another backyard, traveling between the two at high speeds. Undoubtedly, both take-off and landing spots would have to be specially designed to withstand the shock of departure and arrival.

VTOL planes could, in the end, be no larger than automobiles, no more expensive, and as suitable for individual use. They, too, like the automobiles of the future, could be thoroughly computerized and automated. Their advantages over ground vehicles are that the air is three-dimensional and roomier than the surface and that no roads are necessary.

We can imagine the coming of the simplest possible air travel in the form of a reaction engine strapped to an individual harness (something that has already been tried experimentally). One could fly, then, without the insulating effect of surrounding metal and get the actual sensation of flight, which one could not possibly get in any enclosing vehicle, and with much more control than in kite-gliding.

Such personal flight might become one of the great sports of the future but would probably offer only a minor contribution to actual transportation. It would be too slow, and the unprotected human body too fragile, for such travel to be either economical or safe on a large scale.

The ultimate in getting off the surface of the earth is to move beyond the atmosphere altogether in a rocket ship—as has already been done. Human beings have been carried to the moon and back on six separate occasions. This has not brought rocketry to the status of commercial transportation, but the space shuttle is the first step in that direction.

In the course of the next century, it is quite likely that humanity will extend its range permanently into near-space at least and that the earth-moon system will be filled with solar power-stations, with automated industrial plants, with observatories and laboratories, with moon-based mining stations, and, most important, with permanent space-settlements.

Rocket travel will then become the major form of transportation.

Yet every one of these forms of transportation makes use of energy. Even if we have developed our alternative sources to the point where there is no danger of running out of energy, even if we produce our fuels and our electricity in such a way as to preclude noise and chemical pollution, we still face the danger of thermal pollution.

Energy from nuclear fission or fusion, from solar radiation collected in space, from geothermal energy, all add to the energy that the earth's surface collects normally from the sun. This added energy is bound to raise the earth's surface temperature by an amount that though small, would be enough to produce serious climatic disturbances. Even a few degrees of tempera-ture-rise would be enough to initiate the melting of the polar icecaps and the raising of the sea level by two hundred feet—enough to drown the heavily populated coastal lands of the world.

No matter how copious and free energy may be, therefore, we must be careful not to use it too copiously and freely.

One way of saving on energy use would be to do away with unnecessary transportation. For instance, people commute between work and home, or travel long distances to engage in business conferences. With the develop-ment of improved communications and increasing automation, it will become possible in the not-too-distant future, for people to control and maintain business operations and machinery at a distance. The collection of data electronically, the transmission of information by way of image as well as sound, the personal conversation from point to point anywhere in the world, will make it possible for work to be carried on from home, so that the heavy transportation of human beings for business purposes will decrease drastically.

What remains of personal transportation—for social visiting, for tourism, for sport—will become all the more uncrowded, convenient, economical, and pleasant.

---
# 39

# The Corporation of the Future

The corporation is a portion of society, and it follows, then, that if there are all-embracing changes society will be undergoing in the coming decades the corporation will, of necessity, be molded by those changes.

As an example, if humanity is so foolish as to involve itself in a thermonuclear war in the near future, then the changes the corporation will undergo will be such that there would be no need to write this essay at all. Its future will be worth no discussion.

We will assume that this will not happen and that no other catastrophe will take place. An overoptimistic assumption, perhaps, but it is one we will make.

In that case, our safest bet is that the era of high-technology communication, in the beginning of which we are now immersed, will continue to develop and amplify.

We may look forward to more numerous and more versatile communications satellites, laser beams replacing microwaves in space and providing millions of times as many audio and video channels, optical fibers carrying light replacing copper wires carrying electricity, and elaborate computerization making the world more responsive to our needs.

The consequences will be many, startling, and, to a large extent, possibly unforeseeable. (There is no embarrassment in saying this, even though I am a futurist, for all recent history shows that the consequences of advancing technology have always provided surprises to even the wisest and most prudent thinkers.) What is more, they may well be noticeable, and even overwhelming, in a remarkably short period—say, by the millennial year 2000, long a target-year of imaginative writers and now only a few years off.

For instance, it is very likely that we will see an "informalization" of business. It will become less and less possible to delimit a "place of business." There will not be a bloc of space within which the tie is drawn tight and the lips tighter, and outside of which we are free to show a bit of Adam's apple and to smile.

We will be able to dictate and receive our letters from any point; we will talk and listen, display and see, from our homes as well as from anywhere else. Furthermore, information can be received or sent out, in greater volume and with greater speed than is possible now, and from and to anywhere.

It will also become less and less possible to delimit a "time of business," since information, in a computerized world, will be available at all times.

This does not mean that business preoccupation will extend itself to all times and everywhere. Very few people could endure that. It does mean that business decisions will be made in less concentrated fashion and are more likely to be interspersed with periods of other types of activities. And, since the constant, short-notice switch from formal to informal and back again, would be unbearable, the tendency would be to ease the difference between the two, and to increase the informality of doing business.

This tendency would be heightened by the decrease in face-to-face confrontation. Throughout history, it has commonly been necessary for information to be communicated directly from mouth to ear. The use of the written word has made long-distance communication possible (if much slower), and in the past century and a half, the telegraph, the telephone, and radio have successively made long-distance communication more nearly instantaneous. Nothing till now, however, has quite replaced the immediacy and intimacy of mouth-to-ear, and this has meant that, in order to send or receive information, the body carrying the mouth or ear must be sent. Seventy kilograms of mass must therefore be transported in order that essentially massless information be delivered.

With modern communication, sight and sound can be transmitted by closed-circuit television and, with proper communications satellites and the use of holography, it will be possible to conduct meetings between individuals in separate cities (or even separate continents) with all the effect and immediacy of personal contact.

This will enormously reduce the expense of doing business, since it is much cheaper to send an image than a person, and increase the speed and efficiency as well, since the fastest person-carrying transport vehicle could not match the speed of a beam of radiation. It will further encourage the informalization of business, since such image-to-image contact may quite possibly be made from the homes of all participants so that the attitude will be that of host and guest on every side.

All this will lead to an increasing internationalization, even globalization, of the corporation. In a computerized society, with high-technology communications in place, it will be as easy to do business around the globe as within a city.

To be sure, this is already almost true, but, as viewed from the year 2000, the "multinational corporation" of today would seem to be a clumsy and unwieldy phenomenon.

The development of the "global corporation," which will be a natural, and even inevitable, phenomenon, thanks to the nature of forthcoming technological advance, will have interesting consequences of its own.

For one thing, global corporations will serve as a strong force for the development of an international language. Thanks to the accidents of history that led Great Britain to take first place in the exploration and colonization of the non-European world in the eighteenth and nineteenth centuries, and to the development of the (for a time) successful British Empire, English has become the international language of business. This role of English is likely to become even stronger in the future.

After all, English is spoken, as either the first or second language, by more people than any other language in the world, except for Chinese — and Chinese is almost entirely confined to eastern and southeastern Asia, whereas the use of English is distributed widely throughout every continent. Then, too, the United States is the preponderant economic force in the world today and will surely continue to be so for the remainder of the twentieth century. Its closest rivals, Japan and Western Europe, are rich in English-speakers.

The use of an international language is bound to serve as a vehicle for world peace and world cooperation. All history shows that it is perfectly possible to have peoples of virtually identical languages fight bitter wars and civil wars. (There is the example of our own Civil War a century ago, and the apparently insoluble struggle in Northern Ireland today.) Nevertheless, foreigners are bound to seem more foreign and less human if they speak only gibberish, whereas the strange dress and appearance of a member of a far country seems to become muted as soon as he addresses you, faultlessly, in a language you understand.

In fact, we can go beyond language alone. Ever since the development of the nuclear bomb, a full-scale war has become unthinkable. Like it or not, international cooperation becomes absolutely essential as soon as the point is reached where the alternative is a nuclear war. What is more, we have reached a stage of global distress (overpopulation, pollution, declining resources, falling living-standards) such that it is hopeless to expect solutions based on unilateral actions by individual countries. That means that international cooperation becomes absolutely essential as soon as the point is reached where it is clear that problems can no longer be solved at all by anything less than global action.

How can we achieve such international cooperation when, for centuries, suspicion and hostility between neighboring nations has been the overwhelming rule?

It would help to have the example and help of international institutions whose existence overrides national boundaries. Examples of this exist today. The Roman Catholic church is the most familiar and oldest example; but it is strongly ideological and there is sure to be hostility from many of those who are not members of the institution. The scientific community is a global body that is less controversial but it is extraordinarily weak and uninfluential outside its immediate interests.

Global corporations would, without deliberate intention but merely as an inevitable concomitant of attempting to do business in a rational way, produce international cooperation. They would, quite automatically, develop a system of organization that would be suitable for the handling of global problems. By the year 2000, some business thinkers might well be wondering whether the notion of "the withering of the state," which was an article of faith among the early communists, might not take place in a totally different sense. They might look forward to a period in which nations would carry on their ancient business in a purely formal way, being reduced to ritual and pomp, while a consortium of corporations actually handled the details of running the world.

This can't be so without a "responsibilization" of the corporation—a broadening of its responsibilities, that is. The corporation will have to do more than earn money for its stockholders; it will have to assume the task of running the world. (It may well be that it will find it can't fulfill the first function without laboring toward the second.)

Let us consider a few examples of what I mean.

First, it is not very difficult to see that we have come to a major transition point in human history. Physically, the planet is full: more than full, for it is dangerously overpopulated. Moreover, its resources—even very fundamental resources, such as fresh water and food—are being pushed to the limit. We now live in a closed society and, without room for expansion, we might well be doomed.

Fortunately, there *is* room for expansion—into space. Well within reach of humanity today, without any need of breakthroughs in technology, there is an ample, and virtually eternal, supply of energy. We need only build a series of broad sunlight-catching power stations in the equatorial plane capable of turning sunlight into microwaves and beaming them down to Earth. Vast quantities of material resources—metals of every kind, concrete, soil, glass, oxygen—can be obtained by placing a mining station on the moon.

Factories can be set up in space to take advantage of the unusual properties of space—endless vacuum, zero gravity, high and low temperatures, hard radiation. Indeed, even factories doing no more than the duplication of work that can be done on Earth would be useful in space, for undesirable wastes can be discharged into space, where they would be swept away forever by the solar wind. In the end, humanity could have all the advantages of industrialization with few or none of its disadvantages.

In addition, laboratories and observatories can be set up in space and even settlements capable of housing tens of thousands to tens of millions of human beings. This is an undertaking that could be started now, were it not that there are cautious exceptions taken to the expense that would be entailed. It would cost tens of billions of dollars a year. Yet the world spends *hundreds* of billions of dollars a year on competing sets of military machines that can't be used without global suicide.

In a world of global corporations, "responsibilization" would mean the recognition of the necessity of expansion into space. Global cooperation would mean an end to vast military expenditures, and part of the effort could be put into peaceful expansion of the human range. The corporations, cooperating among themselves, could do what the nations today are reluctant to do.

A second example:

With the informalization and decentralization of business, and with the decreasing importance of face-to-face contact, the physical differences between individuals — male or female, white or black — would be less noticeable and would seem to bulk much less in importance. Differences in nationality and culture would also seem less important in a world in which an international language was coming into being.

On the positive side, there would be the fact that, as corporations grew global and took on greater and greater responsibilities, there would be a greater necessity for obtaining the services of those people of drive, ambition, and judgment who would possess the capacity to make decisions correctly and quickly. Such people have never been numerous and, as the decades pass, there will be less and less inclination to skip over some because of a difference in complexion or because of the presence or absence of breasts.

In short, there will be a strong movement to wipe out any remnants of racism and sexism for highly practical reasons. This is not to say that it isn't splendid to be against such malfunctioning of the spirit out of a feeling of justice and humanity, but I suspect it would work better if racism and sexism were recognized as bad for business. That may be an ignoble way of looking at it, but if it means a cure of the disease I will accept it.

It will, after all, be part of the globalization of the corporation to possess a globalization of outlook upon humanity as consisting of a single species with the similarities among its varieties far more marked and important than the differences.

Third, and perhaps most important, will be the responsible attitude of the corporation toward the vast social changes that will soon be brought about by the rapid robotization of industry; of the rapid advance in numbers, complexities, and versatilities of robots; and of their forthcoming advance into the home. (This is something concerning which I have strong feelings, for I myself was the first to consider such problems in detail in a series of science-fiction stories that I began to write forty-three years ago.)

It seems reasonable to suppose that robotic devices of one sort or another—computerized machinery—will replace a large fraction of the workforce.

This is not, in itself, bad. The kind of work a robot can do is work that involves dull and endlessly repetitive operations. To make a human being do this is to stultify him and to place him on the road to imbecility or madness. (This was demonstrated with consummate artistry, once and for all, by Charlie Chaplin in the opening scenes of his classic motion picture *Modern Times*.) The robot, on the other hand, properly designed and properly programmed, can do the work much faster, much better, and with far less harm to himself than ever a human being could do it.

With these factors in consideration, it would seem inevitable that corporations of today, and even more so those of the future, would push hard for the robotization of industry and, generally, of the world.

The problem arises, however, of what one is to do with those workers who are replaced by the robots.

It is not that there will be an overall diminution of jobs. If the past is to be a guide, technological advances create more jobs than they destroy. Thus, the automobile industry employs far more people than the buggy industry ever did. Nevertheless, there is a change in the *kind* of jobs that will be available. The repetitive jobs of the assembly line will tend to disappear. The dull jobs of paper-shuffling and button-pressing will disappear. In their place will be such jobs as computer-programming and robot maintenance.

On the whole, the jobs that will come into existence will be far more creative and will take far more education and training than will those that have disapeared.

It will therefore be part of the responsibility of the corporation of the future to see to the re-education of the workforce. This could be done out of pure feelings of humanity and philanthropy, but it is more practical to suppose that it would be done out of a very natural desire to preserve the stability of society. It might save money, in the short run, simply to cast out the displaced, but it would not be good business to have hordes of hungry and angry people ready to change, by force, the economic system that reduced them to misery.

Naturally, the further question would arise of how one is to deal with those who are too old or, perhaps, mentally unequipped to respond to the kind of training required for jobs more skilled and complex than those at which they had been working. These, presumably, will have to be kept at some level of work they can handle, robots or not; or else be pensioned. There will be a clumsy transition period that will have to be handled with as little pain as possible.

Some might feel, though, that the vast majority of human beings are simply incapable of doing more than dull and repetitive work and that robotization is intrinsically evil because it will forever replace human robots who can do nothing else. Are we to pension off countless millions in every generation?

It is dangerous, however, to accept this possibility without investigation. After all, it is quite possible that it is inefficient education, both at home and in school, that leaves a mind undernourished in the first place. It is possible that it is a mindless job that confirms and extends the malady.

Once the transition period is over, it may well be that, in order to avoid a perpetually renewed situation of this sort, it will be necessary to adopt a radically new approach to education.

Through most of history, education has been left to parents. Youngsters were taught, by precept and example, how to do the work their parents did, whether it was farming, fishing, or cooking. Those few parents who were well off and whose children did not have to work for a living hired tutors to teach those children the classics and to train them in the accumulated culture and philosophy of the past.

The notion that *all* children ought to have an education supplied them by society at the expense of the taxpayer, generally, is a nineteenth century notion. To be sure, that much is necessary in an industrialized society. It is even more necessary as technology progresses, but what was good enough for the nineteenth century and even the twentieth is no longer good enough for the twenty-first. The teaching of children in groups, according to a fixed curriculum, does not allow for the great differences from child to child and is therefore a most inefficient form of education. Most children do not profit by such an education to nearly the degree they might.

What is really needed, if we are to develop a society consisting of a large mass of creative personalities capable of developing the skills necessary to do jobs that cannot be done by robots, is to have each child educated at a speed, in ways, and in directions suitable to that child. Ideally, one teacher should be involved with one child.

How are we to find so many teachers?

The answer lies in the computerized society we are entering so rapidly. The time is coming when every home will have its computer outlet; when every outlet will be connected with a central computerized library that will make available to any person, at need, all the accumulated knowledge of mankind—thoroughly indexed, so that any item or group of items can be called up at will.

In addition to the teaching a child would get from a parent or a profesional instructor, that same child would be free to make use of the computerized "teaching machine" to follow its own interests in the home, at times and in ways and for subjects of its own choosing.

The corporation may then well feel it to be its responsibility (for the sake of developing an educated, efficient, versatile workforce from generation to generation) to take a primary interest in this process of education, and in the details thereof. Each corporation would develop programs that would be particularly useful for developing the kind of trained mind that would be suitable in its organization.

But I have gone far enough to make my main point. I began by saying that the corporation is a portion of society and must therefore be molded by changes in society. I have ended by pointing out that, as technology advances, the needs and duties of the corporation may expand until it is coterminous with society, so that the time may come when it must therefore do its best to mold society as well as be molded by it.

# 40

# The Future of Collecting

Collecting is, essentially, a spare-time activity. It offers an excellent way of filling those leisure moments when we are not required to concentrate on the immediate tasks necessary to promote the security of ourselves and those close to us. Those leisure moments must, after all, be filled in some way, if we are to avoid the painful disorder of boredom.

Collecting, furthermore, is a rather benign activity. Conducted with reasonable care and rational intensity, it gives pleasure to the collector and to viewers, and does harm to none. The time spent on collecting might well be devoted to activities some might feel would be more constructive and socially useful, but it might also be devoted to activities that are harmful and damaging. Since it is very likely that there are many more ways of doing harm than of doing good, to settle on something like collecting, which is, at worst, a neutral activity, is surely to be ahead of the game.

But it follows, of course, that to be a collector one must have some spare time. If every waking moment of one's life is spent in an attempt to wrest a bare subsistence out of a hard world, there is neither time, nor money, nor desire to collect anything but whatever quantity of food, clothing, and shelter one can wring out of existence.

Throughout history, most human beings have been in the unfortunate condition of having to deal only with survival. The very few who have had any leisure to devote — kings and potentates, wealthy landowners, office-holders, and merchants — collected items they valued and could obtain and pay for. They collected wives (or mistresses), jewelry, art, books, and the like.

Such collections were themselves advertisements of the power and wealth of the collector.

Since the Industrial Revolution, an increasing percentage of the population of the industrialized nations has gained leisure time and thus obtained the primary requirement for collecting.

That, however, doesn't make every one of them connoisseurs and patrons of the great artists. All the leisure time in the world will not avail if one lacks the money to pay for those items that are desired for their intrinsic rareness, or for a level of beauty that makes them rare. Even if we imagined a great many people with sufficient wealth to buy art, or diamonds, or first editions, or rare stamps, the effort by so many to do so would simply drive the price upward and leave such things available only to the top echelons.

Where there is mass leisure, then, it becomes necessary to collect objects that are not so generally admired or desired as to be prohibitive in price, but yet possess interest.

A common solution is to become interested in memorabilia or curiosa possessing nostalgic value. The fact that these are not currently in fashion makes them not too easy to find or too familiar to the audience. (The fun is ruined if the game of collecting is made too easy or too commonplace.) Second, they have no intrinsic value, so that the prices are not prohibitive as a matter of course. Third, there are so many different varieties of memorabilia that it is quite possible for a collector to stake out an uncrowded niche as his own, so that competition will not bring about too great an artificial heightening of price. And, fourth, there is a definite social redeeming value to the collection of memorabilia, since these offer information concerning a vanished way of life.

Well, then, what of collecting in the future?

That depends, of course, on what the future is like. If we ruin the planet with thermonuclear war, or by allowing ourselves to run out of energy while we multiply our numbers endlessly, the future may be one of destruction, and of haggard survivors scrabbling in the ruins. In that case, we are back to collecting food, clothing, and shelter.

But that may not happen. If we avoid catastrophe, the present direction of technological change seems to point in the direction of a computerized and automated world, one in which there will be less and less "work" forced on human beings, and more and more time for the development of human creativity, while the dull and repetitive functions that keep the world going are left to machines.

It is possible, in short, that the twentieth century will see a world in which leisure makes up the major portion of the lifetime of the average human being. The most important concern of each human being will then be to fill that leisure time in a way that will be pleasurable to one's self without serious harm to others.

Obviously, one way of passing leisure time pleasurably and nonharmfully, is to engage in the gathering, the cataloging, the display, the maintenance, and all the other duties involved in collecting.

It may well be, therefore, that we are approaching a golden age of collecting.

With the number of collectors increasing, there will be more and more pressure to think of new varieties of objects worth their collection, and the decision on this may depend on the direction taken by society.

We can be well assured, for instance, that in the electronic world of the future we will approach cashlessness as a major characteristic of society. Special cards, computerized to record the fluctuating assets of the owner, can be used to receive and make payments, whether for the purchase or sale of a buttered roll or of a yacht (with taxation making automatic inroads at every step of the game, of course).

This means that the use of coins, of paper money, of checks, and even of noncomputerized credit cards will dwindle steadily, and the objects mentioned will become passé. That will make their collection an entertaining pastime, it may be assumed.

Surely, not all coins and bills will be converted to computerized assets, and there will be many out-of-the-way nooks in which they will be found — and worth more than their face value, too, by a good deal. A dollar bill, in good condition, with no markings on it, would be a proud possession indeed, to be kept in a plastic envelope and displayed to all possible advantage.

I suspect, however, that credit cards will make a more desirable collection. Made of plastic, nonperishable, often colorful, and existent in endless variety — discarded by their original owners or allowed to gather dust, instead of being turned in as assets — they could make a collection that would have a beauty of its own, quite apart from its nostalgic value.

In an electronic world of instant communication, there will be a further blow to personal letter-writing, completing the task that the telephone began.

With closed-circuit television, and instant reproduction of documents by wire, business communications will be totally revised, so that the day of the postage stamp will be over. (That will surely make stamp collections all the more valuable.)

Additional collections of items memorializing a dead world of paper communications might flourish. There could be collections of envelopes of different styles, of postcards, of letterheads. The whole world of twentieth-century stationery could become a challenge to collectors, if only because the items are so frustratingly perishable.

Even more generally, the noncomputerized equipment of a noncomputerized world would become collectible *en masse,* for not only would they serve as interesting examples of design and, in some cases, be workable, but they would also represent nostalgic memories of a time (feared by some, disbelieved by others) when the tools of humanity had no brains at all and could do scarcely anything but operate as the manipulation of human hands forced them to.

To those with space and money, a collection of typewriters might be fascinating, in all their models, electrical and nonelectrical. Each would be capable of working — stupidly, letter by letter, as the keys were pushed. And, in an age when doors opened by voice or by fingerprints, a collection of old-fashioned keys would be delightful and take up comparatively little room.

Any nondigital clock or watch would be of curious interest; in fact, the collector might well have to explain to the viewer just how they could be used to tell the time. What about radio-tubes of different types, or ordinary light-bulbs (both incandescent and fluorescent) in a day when ceilings themselves may be softly luminous?

With the advance of communication and computerization, collections will take on a new role in society.

In the first place, every collection can be itemized in full detail, not only in written description but by three-dimensional holographic image; and the whole can be registered, and the information placed in a computer. The record could be revised periodically; indeed, from day to day, to take into account acquisitions and sales, finds and losses. Collections would thus become very public things, and yet, at the same time, be more secure.

Theft could become very difficult indeed, for anything stolen could not be sold openly without its being recognized at once for what it was when the would-be buyer demanded its registration card and checked it with the computer. If the card was legitimate, the object would at once turn out to be either stolen or a forgery. If the card was not legitimate, the object would turn out to be nonexistent.

If an item were sold to a competing collector who asked no questions, or if it were stolen by the collector himself, then it could not be displayed openly, since this would require registration, which would at once be fatal. Naturally, the person who held an item illegally might find enough satisfaction in viewing it secretly, but surely, in a society where many people would have their collections on display, there would be something unsatisfying in having something one could not put on display.

Computerization, after all, would not be for registration only, but would be an aid to active display when the possibility of theft is so diminished.

Since collections would become so common and so important a way of spending free time, the collector would surely consider his labors to be, in part, a justification of his existence and a demonstration of his ingenuity and creativity. He would have pride not only in the mere collection of the items but in their manner of display and he would be anxious to demonstrate his expertise by showing it widely.

From the viewers' standpoint, there would be as much interest and amusement in a skillful and creative collection as in a night at the theater. One can almost envision contests and blue ribbons. (Has anyone ever made a collection of the assorted ribbons awarded at assorted contests in the world, I wonder?)

Nor would it be necessary to travel the world over to see outstanding collections. These could be viewed by long-distance holographic transmission, in a day when communications satellites and modulated laser-beams could give everyone his own television channel on which to transmit and receive.

In fact, there might well be something intermediate between physical travel to the site of a collection and the necessity of making a home-to-home connection with every collector in whom one might be interested.

There might well be museums that would store holographic records of various collections and which one might visit for special viewing.

The more attractive, unusual, and creative collections might be viewed on a selective basis. Some particular museum might be given (or, much more likely, sold) all rights, and that museum might then charge admission. Or the collection might be leased for fixed periods of time to various museums.

Perhaps no one would view collections with more ardor and attention than other collectors. There is not only a collector's desire to study the content and style of other collections for hints about how to improve or modify his own; but there is also the fact that a viewer who has some object of intrinsic or sentimental or nostalgic interest will know, by careful viewing, who is most likely to make a high bid for it. And, if a collector encounters a collection that in some ways overlaps his own, he might offer exchanges that could strengthen both.

Appraisal and auctioning could be done by holography. Under the improved technology of the twentieth century, holographs will be as revealing and as "real" as the real thing, and the waste of unnecessary transport will be avoided. Once an item is purchased and transferred directly to the new owner, a careful comparison with the holographic record would, of course, be necessary, if only to discourage human wiliness from gaining the upper hand over strict integrity.

Another group of fascinated viewers of all collections would be the sociologists and historians of the twenty-first century, who might very well discover more about the twentieth century from the mute artifacts they study than anything that the printed or recorded word could tell them. How often, one might wonder, would a small, silent object give the unmistakable lie to whole volumes and orations of self-serving nonsense.

One last point. If the twenty-first century sees the establishment of mines on the lunar surface and independent space-settlements in the lunar orbit, there may well be collections of objects related to the human exploration and exploitation of space, collections characteristic of the twenty-first century and of no other.

Then, too, on the various space-settlements swinging in silent orbit about the earth, there may be collections from the home planet; collections that may amuse and thrill those who have spent all their lives on a space settlement and yet still experience an odd twinge of nostalgia, perhaps, for the world of their parents and grandparents.

# 41

# The Computerized World
## Part 1

A generation ago, during World War II, the electronic numerical integrator and computer (ENIAC) was built at the University of Pennsylvania.

It was the wonder of the world; the first fully electronic computer, an "artificial brain." It weighed 30 tons and took up 1,500 square feet of floor space. It contained 19,000 vacuum tubes, used up as much energy as a locomotive, cost three million dollars, and solved problems too complicated for human beings (and did so at enormous speed).

Now one generation has passed; one generation, a little over thirty years.

The rickety, unreliable, energy-guzzling vacuum tubes are gone. They were replaced by solid-state transistors, which, as the years passed, were made smaller and smaller and smaller. Finally, tiny chips of silicon, a quarter-inch square, as thin as paper, daintily touched with traces of impurities here and there, are made into compact little intricacies that are fitted with tiny aluminum wires and joined to make microcomputers.

For three hundred dollars, from any mail-order house, at almost any corner store, one can now get a computer that consumes no more energy than a light bulb, is small enough to be held in the hand, and can do far more than ENIAC, twenty times faster and thousands of times more reliably.

Still, from year to year, these microcomputers grow more flexible, more versatile, and cheaper. Almost anything can now be computerized, so that changing environmental conditions can be taken into account and the workings of a device adjusted instantaneously to suit. Watches, vending machines, pinball games, traffic signals, automobile engines can be outfitted to observe, remember, and respond in a way that will maximize efficiency and adjust to changing conditions.

In the home, microcomputers will be able to respond to ready-made programs that are being devised in ever greater quantity and versatility. Program a microcomputer properly and it will keep track of the Christmas card situation, or do the billing, or organize a tax return, or run the heating and lighting of a house in as complex a pattern as one wishes, judging for itself temperatures and light intensities and adjusting matters to suit. It can water a lawn or turn on a television set, adjust focus, and switch stations in a preset pattern.

To put it briefly, the microcomputer can do anything that is simple and repetitive. It can sense and respond more delicately and more quickly than any human being can; it doesn't get tired, annoyed, or bored; and, properly programmed, it does not make mistakes.

They're catching on, too. As America became TV-saturated in the 1950s, it seems as though it will be microcomputer-saturated in the 1980s.

Is this to be feared?

In 1957, long before computers had been miniaturized and made pocket-sized, I wrote a story called "The Feeling of Power," which dealt with a world of the future in which computers had become so ubiquitous that people had forgotten how to add — until a lowly technician rediscovered the art and astounded the world by announcing that $9 \times 7 = 63$.

Can we really forget in this fashion?

To some extent, perhaps; but let us consider what happened in previous changes of this sort.

When writing was invented, there must have been many who felt that it would spell the end of memory, and that human beings, able to inscribe thought and records, would become witless in their dependence on marks on brick or papyrus.

To an extent, memory did grow less important, but it didn't vanish, and who now would give up writing just so human beings would have a chance to exercise their memories (and argue with each other when memories failed to agree).

Think of the invention of the yardstick or the sundial and how that limited the necessity to learn to make clever estimates of distance and time. Those talents would go to waste, it seemed, since any fool could read a yardstick or a sundial. But who wants to go back to estimates, no matter how dainty a talent that might be?

Who wants to give up printing because we have lost the beautiful illuminated manuscript; or the typewriter because we have lost Spencerian penmanship; or the telephone because we have lost the genteel art of the "at home"?

These technological advances do cost us some of what we had before, but never all; and what they put in its place is much greater than what they cost us.

Advancing technology has indeed mass-produced dime-store gimcracks and lost us the lovely handmade artisanry of yore—but that artisanry was for a small population of the well-to-do who could afford it, and this small population would include scarcely any of those who are reading this article now if they lived a few centuries ago. (It would certainly not include me.)

Technology gives a touch of comfort to hundreds of millions who would never have had it otherwise and who can scarcely be expected to bewail the lost specialties of the rich.

Will we become too dependent on our computers, though, while they continue technology's task of bringing comfort to us all? If the computers stop, will we then find ourselves unable to survive?

We'll become dependent, of course, and will find survival difficult. In fact, we needn't use the future tense. We are already, right now, dependent upon computers, and would find it difficult to get on without them. Stop all the computers and every major industry would stumble and trip because of an inability to carry out its paperwork, to say nothing of a thousand different industrial policies.

Our armed forces would find themselves helpless; scientific research would limp; and worst of all the government itself would be paralyzed, if only because the IRS would be instantly out of business. (And don't say "Good riddance." The United States can survive the death of a president or the paralysis of Congress or any disaster you can name short of a thermonuclear war; but, make it impossible for the nation to collect its taxes and process its payments and we'll have uncorrectable chaos in a surprisingly short time.)

Is the dependence on computers—great now and rapidly growing greater—a disaster? Perhaps not.

We were all utterly dependent on a variety of complexities in our society before computers were ever invented. We depend on the intactness of the wires that fill our cities with electricity and on the generators that keep them humming. We depend on the pipes that carry our water-supply and on the transportation facilities that bring food in and garbage out, on the fuel that supplies us with heat, on any number of things.

We can lower the dependence, yes, by unwinding all the progress human beings have made in centuries. We might somehow make three-quarters of the world's population vanish, shrink the cities to villages, dismantle all the complexities of industry, and become a planet of farmers—and then we will all be dependent on the health of our horses, herds, and flocks, on the supply of firewood, on the coming of rain.

And we can't voluntarily go back. All through history, human beings have turned away from simplicity and adopted complexity whenever they had a chance. They chose those dependencies that, while they lasted, made life richer and more comfortable, and turned against those that, even while they lasted, broke backs and wore out bodies.

In line with this the world will continue to move forward toward computerization while it can.

# Part 2

With every important revolution in technology, the role of human beings in the economy alters and some varieties of work ceases to be.

There was a time when 95 percent of humanity dug in the soil, herded animals, gouged out minerals, or sailed on ships to obtain the food and other raw materials needed by themselves and the remaining 5 percent.

The Industrial Revolution, which began two hundred years ago, increasingly shifted the bulk of this labor from human and animal muscle to the machine. The need for what we now call "unskilled labor," the sheer straining of sinew, declined, and the need for skilled labor and for services increased. Skill meant the need for mass education. Leisure meant the need for mass entertainment.

The coming of the computer will require further shifts. Unskilled *mental* labor will be on the way out. More and more, the dull processes of shuffling and adjusting and checking and listing, and all the other things that first irritate and then stultify the brain, will be done by computer.

And what will be left? Leisure. Amusement. Creation.

We're getting a taste of it already. One popular aspect of the microcomputer in the home is its ability to be programmed as an adversary, its ability to play games with human beings.

Computers can be programmed, for instance, to play chess. While chess has never been completely analyzed, and may never be, a program can guide the computer according to some general principles and make it possible for it to play a passable game.

If a person learns the rules of the moves, he or she can begin to play at once and, of course, be soundly trounced by the computer. The human player can, however, learn from his own mistakes (that's the beauty of the complex programming of the human brain) and improve his game. And he will learn more effectively than against a human adversary.

A computer adversary, after all, does not get tired, or impatient, or contemptuous, or busy with other things. It can be used at will and at the human player's own pace; and eventually the human will learn to win.

He or she can then buy a better program or, for that matter, construct one, or seek out human adversaries, or find a different game.

He can write music and have the computer play it back, or he can construct a program what will make it possible for the computer to devise plots or write poetry, which the human partner can then use as a springboard for creative improvement. Or he can have the computer simulate houseplans on a television screen and play the game of interior decoration.

To be sure, there is a streak of Puritanism in many of us that would lead us to disapprove of this sort of thing as "playing" or "fooling around." There is the fear that dependence on the computer for our amusement will cause our own mental abilities and self-reliance to go slack and rot.

That, however, may be precisely the wrong way of looking at it.

It is the unrewarding and repetitive scut-work of today, occupying, as it does, only the surface of the mind, that rots our mental abilities, and it is the creative "play" that can enhance and stimulate them.

It is possible, in fact, that the coming of this new "play" that computerization makes possible will be but the very small tip of a huge iceberg and that the new age of leisure is the route to new advance.

Until now we have labored merely to maintain our social and economic structure, and there has been very little time and energy left over to advance it. With full computerization, the world will run itself with only minimal human supervision, and the major part of human thought and energy can be put to extending and intensifying the structure of society.

Then, too, a leisure culture may, in itself, lower the birth rate—at least, leisure has always seemed to have had that affect—so that computerization may be an important step toward solving the population problem (though it probably won't work quickly enough for us to be able to depend on it alone in this respect).

Furthermore, it is a small step from computers that play games to computers that educate. If we can learn to play chess by using a computer, could we not do the same where the computer made other information available, and might not a computer possibly have the knowledge of humanity at its disposal?

As computers grow more compact and versatile, we can, without difficulty, foresee a day when libraries can be microfilmed and computerized; when all their contents can be carefully classified and any individual item retrieved on demand.

It isn't hard to imagine ourselves asking for the population of Ascunción, Paraguay, and having a computer sort through population statistics in some central data bank and giving us the answer. We can ask more subtle questions and obtain references to textbooks, research papers, monographs, popularizations. We can have particular references reproduced on a television screen or transcribed on paper and, having investigated what we receive, we can ask the question again in a more refined and detailed way, or move on to subsidiary and tangential questions.

After all, in order for civilization to survive, the birth rate will have to drop and the age pattern of humanity will continue its present shift in favor of a larger percentage of mature individuals and a lower percentage of young ones. If we are to prevent the older segments of the population from being a dead weight that will crush the diminishing base of youth and innovation, we must make education a universal opportunity for all and not for young people alone.

The computerized library and teaching machine will make it possible for anyone at any age to investigate anything and go as far as he or she likes in any intellectual direction, whether deeply or trivially, either intensely or dilettantishly.

This could result in a world in which the general level of intellectual curiosity and liveliness would be greatly enhanced and in which people of any age would have the mental sprightliness we tend to associate with only the young now.

Will the preoccupation with computers, whether for games or for solid education, produce a society of isolates who will forget how to talk to human beings? I described such a society in my novel *The Naked Sun,* but it doesn't have to be so.

It is quite possible, after all, for people to lose themselves now in books, in record-players, in television. Computers would offer nothing new in this respect.

Remember, too, that there is another side to the coin. There are thousands of people so fascinated by the television program "Star Trek," for instance, that it begins to fill a substantial portion of their lives, yet it does not necessarily isolate them. Instead, it can drive them to seek out others like themselves, to form fan groups, to hold fan conventions, and so on.

In short, what seems at first to be a force for isolation can become a pull toward human interaction.

To play games with a computer may drive one to test one's skill against other human beings; to learn by means of a computer may drive one to try to educate others.

Imagine a world in which no two people move in quite the same computerized-educational direction and almost all are afflicted with at least some missionary zeal. We could have an intellectual ferment such as the world has never seen.

# Part 3

Will computerization of the world merely affect the surface of society, just liven things up a bit and make life more intellectually stimulating?

Actually, it could make it possible for human knowledge to take enormous leaps forward. Consider this, for instance—

Human knowledge has gained most where simple problems are involved. In astronomy, we deal in large part with matter that we can consider as simple points, moving under the influence of a gravitational force in a way that can be described in a simple equation. In physics, we deal with moving bodies and with other forms of energy that can also be described in fairly simple equations. Chemistry is a little more complicated, but can still be handled.

There is enough of the simple in the movements of stars, planets, billiard balls, and atoms to make those who deal with the physical sciences look pretty good.

What of those, though, who try to deal with more complicated systems? What of biologists who try to deal with the complex behavior of molecules in living tissue, and with the behavior of organisms in evolution and in social structures? Psychologists, who must deal with the human brain, the most complex structure we know, are worse off still; and sociologists and economists, who must deal with human societies, are even further in the mire.

It is not surprising that the social sciences are so badly off in comparison with the physical sciences, and that social advance seems to lag so far behind technological advance as to make modern technology a potential death-trap for our still primitive societies.

Nor will things ever improve as long as we have no tool to aid the mind that is better than those we have had before computers came along.

221

As better and more elaborate computers are developed, it should be possible to solve ever more complicated problems ever more quickly. (This is not to say that every problem is exactly and generally soluble, even by the best computer. Even then, however, *approximate* answers can be obtained and approximations may be sufficient for immediate purposes.)

The computers to come may aid in the determination of technological side-effects. It often happens that some technological change that looks perfectly useful turns out to have unexpected side-effects that could prove exceedingly harmful. A dam, much needed for irrigation, may irretrievably damage the ecology of a region in other ways; a certain method of fertilizing the land may introduce a long-term deterioration of the soil; and so on.

These things can't always be easily predicted, but if we can work out relationships that govern the various items in the picture, we might allow a computer to produce a simulation that it can then follow through time.

Already the Club of Rome has been using computers to try to predict the future of society, given certain changes in population growth, pollution, resource deterioration, and so on, under the influence of different courses of action. The results have been much disputed, because the value of the original assumptions and the nature of the programming have been under dispute.

As we refine our knowledge and our programs and make use of computers capable of handling more variables, the results are liable to become increasingly useful.

Until now, sociology has not been an experimental science. It couldn't be. There is no way of setting up societies of human beings (as we do of rats) in order to subject them to various stresses and note human reactions, human delights and miseries, human lives and deaths, and draw conclusions from it all. Human societies are too complex, human lives are too long, and human rights are not to be tampered with in this fashion.

But we can set up the societies in terms of symbols within the transistor-network of a computer. We can learn enough about societies (with the aid of a computer) to know how to set up our assumptions and relationships, and we can allow the computer to calculate the development and predict its course.

In this way, sociology, and the other disciplines that are now too complicated to handle, may become true sciences.

To be sure, the time may come when we will have sociological experiments of a more direct kind. We may have space settlements some day.

The physical properties and components that will make such a settlement habitable can be easily calculated, but what of the society that will occupy it?

The settlement may contain anywhere from ten thousand to ten million people, and each settlement may draw people from a different mix of cultures, and with different blueprints in their collective minds as to what it is they want to do and how it is they want to live.

How will each blueprint work?

The easiest procedure might be to let each take its chance. Those that prove unworkable for any reason will drop out; the settlement will be abandoned or self-destroyed. Others might prove surprisingly workable, and yet attempts at duplicating them under slightly different conditions (after all, the number of variables from settlement to settlement would be enormous) might fail.

This would be a simple way of distinguishing those that work from those that don't—the hit-and-miss survival of the fittest—but how fearfully expensive in money and in human misery.

The day might come when no settlement would be established without a computer simulation of the planned society. Settlers might choose to go ahead with their plans even if the computer prediction was an unfavorable one. But then at least they would know what they were warned against and, if the reality begins turning in the predicted direction, they could labor to correct that, or could abandon the settlement before catastrophe strikes.

It may also be that a settlement works out well (or ill) in an unpredicted form or direction, and then the very difference between reality and simulation will help improve sociological theory.

Naturally, it is only too easy to suppose that a computerized sociology will teach tyrants how to be more effectively tyrannical and that the computer may help fasten new and more durable chains on humanity.

There is no way of guaranteeing that this will not happen. It may be, though, that a computerized sociology might indicate that a tyrant might best endure if not too tyrannical; and if some freedom is granted in consequence, that freedom may turn out to be contagious.

After all, tyranny is as old as humanity and has got along so well without computers that we can scarcely fear worse with them.

We can argue that tyranny arises out of fear. The tyrant, seeing dangers and hatred in every corner and only dimly aware of the workings of society, preserves his rule and his security by striking arbitrarily in every direction, harder and harder.

It is at least possible that, if he were guided by a computerized sociology, he would learn the true direction from which danger would come and he could afford to be only selectively tyrannical. He might even learn that the true source of his danger was his own tyranny.

I don't guarantee that this is how it will work, but it might. Since nothing else has worked and, since through all human history freedom and the respect for human rights has been notable for its absence almost everywhere and at almost all times, computerizing the world might offer us little to lose and possibly much to gain.

# Part 4

It is odd that science-fiction writers rarely, if ever, dreamed of computers until computers appeared on the scene.

On the other hand, manlike, thinking robots have been written about frequently for over fifty years now, though none has yet truly appeared and none are on the immediate horizon.

The reason for this, very likely, is that it is hard for people to imagine intelligence except in human shape. As long as a computer is a box, little or big, and communicates by a keyboard or by flashing signals, its intelligence remains suspect. Place the computer in a cavity about the size of a skull, equip it with a vaguely human body and limbs and voice, and we will find it hard to believe that it isn't intelligent.

Is it at all likely that the time will come when we will have robots? Can we fit a computer into a skull-sized cavity?

Yes, of course, and there is room in such a space for a fairly elaborate computer, too, given present-day techniques.

Can such a computer compare in any way with the human computer we call a brain?

Ah, that, no. It doesn't even begin to. Even simple computers can do some things, such as repeated additions and subtractions, trillions of times more rapidly and surely than human brains can; but addition and subtraction are very small parts of the talents of the human brain.

It is only with difficulty that a computer can be made to read, and then only if carefully formed letters are prepared for it. The brain, however, can easily read a wide variety of prints and handwritings and is scarcely slowed by misshapen letters and misspelled words—and this, too, is but a small part of the capacity of the human brain.

The human brain has some ten billion neurons, or nerve cells, and perhaps ninety billion auxiliary cells, all of them interconnected in an enormously complicated pattern that we don't yet understand. Each neuron, which is itself far smaller than any unit we can build into a computer, is not merely a flip-flop, on-off mechanism, as are our microtransistors, but is composed of vastly complicated arrays of enormously complicated molecules of uncertain but possibly enormous capacities.

Does this mean we can never hope to have a useful robot?

Well, first, a robot need not be as complex and as intelligent as a human being. If needed to perform certain tasks, repetitive and simple, the robot's computer need not be one capable of translating Chinese into English, or of composing a symphony. We can have very stupid robots, capable of no more than understanding and following a specific list of simple orders — but that might be good enough.

The greater difficulty, in fact, may be that of supplying the robot with a reasonably compact and long-lasting power supply, one that is easily renewed, and of equipping it with the necessary mechanisms that will allow it to move its parts in ways that will make it possible to fulfill its functions.

Second, computers have developed at an enormous rate in the past thirty years and there is every indication that they will continue to develop at this rate for some time to come. Even if robots are very stupid at first, they may well become rapidly more intelligent as newer models are developed in rapid succession. (My book *I, Robot* presents a picture of this.)

Third, it is not absolutely essential that each robot carry its own brain, any more than it is necessary for each television set to incorporate its own broadcasting station. The computer that makes it possible for robots to sense orders or conditions and allows it to answer and to perform its function may be located elsewhere and be connected with the robot by some radiative mechanism.

In such a case, there would be no severe size-constraint on the computer. We might even imagine a computer so large and complex as to be capable of serving all the robots in a city. (Yes, if something went wrong with such a computer, all the robots might shut down at once — but if something goes wrong with a central electrical generator, all of a large city may be blacked out. It's something we have to live with.)

But why bother with a humanoid shape? Would it not be more sensible to devise a specialized machine to perform a particular task without asking it to take on all the inefficiencies involved in arms, legs, and a torso. We might design a robot capable of holding a finger in the air to test its temperature and of then turning a heating unit on and off in order to maintain that temperature nearly constant. Surely a simple thermostat made of a bimetallic strip would do the job as well.

Over thousands of years of human civilization, however, we have built a technology geared to the human shape. The height and form of all human products intended for use are designed with the thought in mind of how the human body bends and how long, wide, and heavy the various bending parts are. Tools and machines are designed with controlling parts that fit the human reach and are adjusted to the width and position of human fingers.

Think of the problems human beings have if they happen to be a little taller or shorter than the norm — or even just left-handed — and you will see how important a good fit into our technology is.

If we want a versatile controlling-tool, then, that can make use of human devices and that can fit into human technology, we would find it useful to make it in the human shape, with all the bends and turns of which the human body is capable, not too heavy and not too abnormally proportioned.

But if computers become steadily more versatile and complex, does this not mean they may become disconcertingly intelligent?

Yes, undoubtedly. The human brain is made up, as far as we know, of nothing but matter, energy, and enormously complex organization. If computers can be given components small enough, and with those components arranged complexly enough, there is no theoretical reason why something comparable to human intelligence can't be produced. (How long this would take is not easy to predict, of course.)

And if a computer can be built to be as intelligent as a human being, why can't it be made *more* intelligent as well?

Why not, indeed? Maybe that's what evolution is all about. Over the space of more than three billion years, hit-and-miss development of the organization of atoms and molecules has finally produced, through glacially slow improvement, a species intelligent enough to take the next step in a matter of centuries, or even decades. Then things will *really* move.

Will human beings refuse to produce a kind of intelligence that will surpass their own? Can we be sure of having a choice in the matter?

Suppose new computer designs are worked out by computers, so that, in a sense, computer-evolution becomes computer-directed. Might it not develop with explosive speed and beyond our control?

But if computers become more intelligent than human beings, might they not replace us?

Perhaps not. Computers are developing intelligence from a different direction, by different steps, out of fundamentally different parts, and for different reasons from those the brain did. Even if computer intelligence surpasses our own (as in some exceedingly trivial ways it already does), it may remain behind in other ways.

The two intelligences, human and computer, may supplement far more than compete and, in cooperation, may do far more than either separately could.

Perhaps, when the time comes that Earth-spawned intelligence emerges from the birth-planet to make whole sections of our galaxy its home, it will be two intelligences that do so. And the senior partner, the human brain, may still have capacities not yet fully understood or fully duplicated that will remain essential to the partnership.

---

## 42

# The Individualism to Come

We live in an age of mass communication, and it is that which makes it possible for a political unit to be enormous in area and population and yet, at the same time, be a democracy.

In the long ages during which the only way instant communication was possible was by a carrying voice, a political unit could allow individual participation in the government (the essence of democracy) only if it were small, like the ancient Greek city-state of Athens.

If a political unit was large, like the ancient Persian Empire, and the ruler could make his voice heard to only his chief officials, then the result was an autocracy. If the political unit had begun small, as in the case of Rome, a decrease in individual freedom with growth was inevitable.

When means of communication become more sophisticated, whether through something as relatively primitive as more and better roads or something as advanced as television, then individual participation in the machinery of government can increase.

*Can* increase, but it doesn't have to. When mass communication is kept *only* mass — such as when the voice of the one leader is all that is heard throughout the nation, thanks to electronics — then the ideal can become that of making the nation a mass automaton, obedient to the single voice. The most dramatic case of this was in Nazi Germany.

Clearly, the situation most favorable to democracy and to individual freedom is where communications exist in forms that can be both mass and individual. Either alone is insufficient and even dangerous.

Radio, television, newspapers, magazines, even billboards and posters, are essentially mass-communication devices. Whether we are speaking of a speech by the president being broadcast on all television and radio stations or of an editorial or advertisement in a small town weekly, we are dealing with communications thrown out indiscriminately. There are no names attached to the receiving end, and all receivers get the same message. The individual is submerged and, in every case, the intention must be to please as many as possible and to disregard those few who don't fit into the mass.

At the other extreme, the most individual form of communication remains word-of-mouth conversation. If you have a point to make, a warning to give, an item to sell, you can speak to person after person, carefully adjusting your words to the needs and personality of each listener. This has its value, but your efforts are, of necessity, extremely localized.

The one form of communication we have, at present, that is both mass and individual, and extremely effective both ways, is the mail. Hundreds of thousands of identical items can be placed in the mail, all of them reaching their destination in a very few days; and this can fairly be considered a form of mass communication. However, each individual piece of mail is addressed and delivered to one particular person among all the people on Earth. So it is individual communication as well.

The individual aspect of the mail would of course be ruined if copies of the same item were poured into the system for delivery to every individual who could be reached, without exception. If that were the case, then the utter lack of discrimination would make the mail no improvement in this respect over the electronic media — and much slower.

The whole point of the use of the mails, then, lies in the possibility of selectivity. A drugstore opening in Fargo, North Dakota, doesn't mail its grand opening announcement to homes in Columbia, South Carolina. An airline, advertising reduced fees on round-trips to California, doesn't mail its news to Los Angeles.

There is an obvious desire by everyone making use of direct mail to achieve his purposes (whether to sell articles, raise funds, inform, plead, or warn) more efficiently and less expensively. We can expect, then, that a major effort will be made to make mailing lists more selective and therefore more useful.

The key here is the computer, which can swallow information and then disgorge it on demand. The raw information would be names and addresses, together with specific qualifications and information concerning each. On demand, a printout would include only those names and addresses that fulfill certain qualifications — such as a list that includes all college graduates with English majors living in towns with populations of between fifty thousand and a hundred and fifty thousand; or all owners of apple orchards in the states of New York and Wisconsin.

The inevitable trend would be toward the establishment of a national computer-bank, government-run (inevitably), from which any kind of mailing list could be obtained. Indeed, the population will eventually resign itself to being coded, with symbols representing age, income, education, housing, family size, and anything else that might be rationally categorized—and having these symbols periodically brought up to date. An attempt to evade or falsify such symbols would clearly be an anti-social act and would be treated and punished as such.

The thought may be revolting to us today, but the trend is irresistible. As the complexity of our society increases, the quantity of our number-designations increases. There are now area codes, zip codes, and Social Security numbers, all of them unknown forty years ago, and all absolutely necessary now if the telephone network, the post office, and the tax system are to work.

Despite the romantic objections of the old-fashioned, we can't abandon these numbered codifications. Indeed, we will surely need more of them.

We may cry out, "I'm a person, not a number," but that is useless. Indeed, we are persons only to those who know us personally. To everyone else, we are, at best, so many names, and a person's name is as much a conglomeration of meaningless sounds as a number is. In fact, since a number is more easily dealt with *en masse,* you are more apt not to be neglected when you are a number than when you are a name.

Wouldn't such coding be an invasion of privacy? Yes, of course, but we live in a society where privacy is impossible anyway. We have chosen to demand of our government all sorts of services (including, chiefly, what we conceive to be military security) that cost so much money that an enormous tax system *must* be set up. In order to make sure that the necessary tax money is collected, the government has to know all about us and must, for instance, be able to look at every check we write. By encoding ourselves thoroughly, we could make this unavoidable snooping more efficient, less expensive, and, we can hope, less obvious and annoying.

But will not all this personal snooping enable the government to control us more ruthlessly? Is it compatible with democracy?

The truth is that no government is ever at a loss for methods to control its population. The history of mankind is a history of tyranny and of government by repression. The most liberal and gentle government will quickly turn repressive when an emergency arises. And this trend is encouraged by the *lack* of knowledge concerning the population.

In the absence of detailed knowledge about its population, a government can only feel safe if it represses *everybody.* In the absence of knowledge, a government *must* play it safe, *must* react to rumors and suppositions, and *must* strike hard at everybody, lest it be struck. The worst tyrannies are the tyrannies of fearful men.

If a government knows its population thoroughly, it need not fear aimlessly; it will know *whom* to fear. There will be repression, certainly, since all governments will repress those it considers dangerous; but the repression will not need to be as general, as enduring, or as forceful. In other words, because there will be less fear at the top, there will be more freedom below.

Furthermore, a thorough knowledge of the characteristics of its population can make the government more efficient in providing those services we now demand. We cannot expect the government to act intelligently if it does not, at any time, know what it is doing, or what, in detail, is demanded of it. We must buy service with money in the first place, as all taxpayers know; but we must then buy useful and efficient service by paying out, in return, information about ourselves.

This is not even something new. The decennial census has grown steadily more complex with the years, to the benefit of the businessman and the administrator, who find in it the information that can help guide their responses. The encoding and computerization of the population is the final elaboration of a kind of continuing and enormously complex census.

The technique for using the central-computerized-continuous census will very likely first be developed in detail by the direct-mail advertisers, since for them there will be a powerful and immediate profit motive. The more efficiently they select a group from the entire population, the higher the return on their expenditure.

The optimally efficient subgroup to be selected as the mailing list must be not so tight as to cut off too many fringe sections with sizable quantities of potential customers, nor so broad as to bring in an outer fringe with so few potential customers as to be an uneconomical addition. Methods for doing this may be enormously complicated, and we can foresee a new and growing specialty, a kind of amalgam of psychology, economics, and mathematics—let's call it "population analysis." The techniques involved will probably be practical only through computerization.

The service of all this population analysis to the seller is obviously tremendous, but it is enormous to the potential customer as well. To eliminate a person from those mailings to which his chance of response is less than a certain level is to free him from precisely that aspect of direct-mail advertising that is most likely to irritate him. He will not be plagued with things that seem to him to be of no interest. As a result of these improved techniques, each person will receive either less direct-mail advertising or that which is more pertinent, or both. In each case, the recipient will be either less burdened or more interested, or both.

In fact, so increasingly narrow and accurate will be the target that each recipient will begin to feel himself all the more an individual through the direct-mail advertising he gets. What he receives will be so likely to be of interest to him and to be slanted to his particular needs that, even if he does not buy, he will feel that someone has gone to the trouble of knowing what he might want.

It is precisely because the society is computerized that he can be made to feel so individual because he is so well-targeted. It is in the noncomputerized society that his wants and needs are unknown to anyone but himself and his immediate associates and he becomes a faceless nothing.

The techniques for subgrouping the population efficiently, which is most likely to be developed first through the necessities of competitive advertising, can then be used for other aspects of direct mail as well.

Organizations that need free-will contributions, political or religious groups that wish to persuade the public to their views, any group that wishes to educate the public about anything from population planning to signs of breast cancer, any group that wishes to warn the public against anything from forest fires to littering, can multiply their effectiveness by the proper selection of mailing lists.

The future of direct-mail advertising is, of course, also bound up with the future of postal service, and there we have the influence of electronic transmission.

Essentially, the postal service delivers information from one point to another, but it is not the information itself that taxes its physical equipment. It is, rather, the paper that carries the information that poses the problems.

To deliver the actual sheets on which the information is placed, and the actual envelopes that encloses them, requires the movement of various vehicles, some as fast as an airplane, some as slow as a walking mailman. The result is that it takes days to make delivery.

The information itself, however, could streak through space at the speed of light (186,282 miles per second) by way of an electric current or a beam of radio waves and be delivered from one post office to another in virtually no time at all.

It is within the range of technical possibility now to scan a printed page at one place and reproduce it at another thousands of miles away. Only the information will have been transmitted, in very little time and with the expenditure of very little energy.

There are, of course, considerations other than time and energy. There is, for one thing, privacy. Someone writing a personal letter may place a premium value on just that. The electronic transmission of information will, after all, expose the contents of a letter to unauthorized eyes (at least, despite all precautions, it will seem so in the thoughts of many letter-writers). The post office will therefore have to offer the service of delivering sealed letters nonelectronically for those who wish it done—probably at premium rates.

(Considering how many postcards are written, and how many more would be if postcards were as socially acceptable as sealed letters, it may be that the amount of privacy demanded will be less than we think.)

It could be that some direct-mail advertising, too, would be more effective if kept sealed, but we can be reasonably sure that almost all of it could be transmitted electronically over long distances. A master copy could be transmitted at one electronic-postal station and the desired number of copies could be reproduced at various receiving points, each copy bearing an address printed out from the mailing list obtained from the central computer.

Physical delivery need only be local then.

The reproduction of catalogs might represent a greater difficulty, but the old-fashioned catalog, listing everything a supplier can possibly offer, is bound to be outmoded once the development of computerized lists is properly refined. Some catalogs may now have hundreds of pages of baby clothes and other infant needs that are useless and (human nature being what it is) irritating to childless couples, or hundreds of pages devoted to large appliances that are useless and irritating to people living in two-room apartments.

There will not be one large catalog, then, but dozens of different small catalogs, each designed for a particular mailing list. Each would be easier to reproduce and easier to deliver. Each would refrain from annoying the potential customer with lists of grossly useless items. And each would make that same potential customer feel more like a highly regarded individual by showing itself, on every page, to be geared to his or her particular needs.

We can go farther than that, too, and should. After all, electronic transmission from city to city displaces only the easiest step of the shifting of bulk-mail. The real time-consumer is the local transfer along country roads and through city traffic-jams. Can we hope to bypass that?

Yes, it is conceivable that we can. We are already in the age of the communications satellite and the laser beam, and both are constantly being improved in terms of subtlety and versatility.

The communications satellite is distance-insensitive. A beam of radiation sent out to a satellite can be amplified and relayed to any point on Earth (via one or two other satellites, if necessary). A message can be sent from New York to Los Angeles via satellite as quickly and as cheaply as from New York to Hoboken.

A laser-beam carrier in visible light or ultraviolet has the potentiality of carrying many millions of telephone, radio, and television channels. With millions of channels available, it would become possible to divide the United States into areas within which each person could have a particular television channel assigned to him. Receiving communication by television would then become as routine and as personal as receiving it by telephone today.

A great deal of the work of the post office would then be to regulate the use of these personal television channels. Much of the information now sent by mail could be sent through the air on the personal channel, to be viewed in the home or to be printed out for a more or less permanent record.

(Since television channels can be tapped, private mail, laboriously transmitted by vehicle and by man, will still exist in some relatively small amounts; but again most direct-mail advertising need not be private.)

When each individual has his personal television channel, a large percentage of direct-mail advertising will be "mailed" in this fashion. It will not be ordinary television-advertising, since it will not be broadcast indiscriminately. And each channel will be included in the individual encoding within the computerized "census."

The selection of the list (by computer) and the delivery of the message (by laser beam) will be as nearly instantaneous as it can be expected to be. It is not to be supposed, of course, that a particular personal channel will be at all times activated, or that it might not be in actual use at the moment it is wanted. Material will therefore as a matter of course have to be stored electronically as it arrives.

Very likely there will be a signal light to indicate that a message is waiting to be viewed. When the personal channel is then activated, each item stored will be displayed in turn. Each can be scanned and erased, scanned and temporarily returned to storage, or scanned and printed out, after which the next item would appear. It will be very much like going through one's mail today, with its mixture of personal items and advertising, in which some are discarded, some put aside, and some filed.

For such direct viewing, single-page items may be best. For multi-page items, it might be preferred to prepare what we would today refer to as "cassettes." By the time personal channels are present in every home, such cassettes should be developed to the point where they are considerably less bulky than the printed material they represent. They might be objects, no more than postcard-size, carrying dots of microfilm.

Such cassettes, delivered by "old-fashioned" methods of mail delivery (but more efficiently, we hope, since there would be so little competition from other items using that type of mail) would be inserted into some appropriate attachment on the television set and played through on the personal channel.

A catalog could be played through page by page with controls that would hasten it on, or back it up, or freeze a particular page.

Communication would be two-way, too. A potential customer, feeling the need of more information than that given in the catalog, could send out an inquiry coded to the personal channel assigned to the supplier. Such messages, sent out from individual homes, would be stored on the supplier's much larger and more elaborate receiver and eventually (not too eventually, we can hope) be answered.

Items could be ordered in this way directly from the catalog, by sending out the code number of the item, the number desired, the individual's own code number, the code number of his credit reserve—numbers, numbers, numbers.

(The exact amount in your credit reserve would be available to you at any time. Any payment made to you for any purpose would be entered and added; any expenditure would be subtracted with a note as to its tax-deductibility. At periodic intervals, the various governments who lay claim to you will take their respective cuts. Ideally, it would be a cashless society we will be living in—but that's another story.)

For that matter, the cassettes need not be a matter of reproducing the printed word only. The pictures can be made to move, and there can be a synchronized recorded message.

The message itself would, again, be keyed to *you,* since it would be designed to suit you as a member of a particular list. The local manner of speech would be used at a level suited to your educational background (or slightly higher if the final decision of the psycho-economist is that a somewhat—but not *too*—superior manner of speech would be effective).

Here, too, where the business advertiser makes his pioneering way forward, other types of direct video-mail appeals will follow.

The personal television-channel may make it necessary for political leaders on the national level to abandon the standard television speech to the nation, carefully geared to the kind of smooth fare guaranteed to offend very few. Speeches will have to be delivered in hundreds of versions, by hundreds of surrogates, each zooming in to the homes of some appropriate mailing list, each speaking particularly to the regional, economic, and educational level of that particular group and addressing himself to its needs, concerns, and interests.

This is done now, in a clumsy way, when a president addresses different groups at different times. It will become more efficient as the lists become more specific.

Will this make it possible for cynical hypocrisy to reach new heights? Perhaps—if the government alone was able to make use of this advanced version of "direct mail."

It would, however, be the essence of democracy that such routes to the population be open to anyone. The opposition could do the same. Any pressure group could do the same to the limits of its ability and ingenuity, just as they can all now use the mails.

If political views (or any other ideas being sold on the open market of the intellect) are broken up in such a way as to aim at as many lists as possible, and if each list is dealt with as skillfully as possible, there remains this to be considered—

The public will be getting specific statements addressed to a number of small and homogeneous audiences. This would force leaders to consider the particular needs, concerns, and interests of *each* audience far more intensely and clearly than, in many cases, they have up to now. And contradictory statements, rousing contradictory hopes for short-term political advantage, would fail drastically in the long run if each group remembered very well the specific statements made to itself specifically.

Furthermore, each subdivision of the population thus addressed would surely have a greater feeling of having the attention of the government, and a fuller sense of participating in the government. Even when one group has to be told it must lose out because of the more pressing claims of another group, the personal nature of the communication would make it easier to take. The heightened sense of individuality that would result would, we might hope, reduce alienation and produce a greater sense of national purpose.

But let us get back to direct-mail advertising. However efficiently information might be transferred, the time will come when a physical object, not just information, must be transferred. Whether it be an egg-beater or a grand piano, it cannot be delivered electronically, at least not by any reasonably foreseeable technique.

How will such delivery — call it parcel post — be handled in the future? All the delivery routes and methods now available will remain and, we may hope, will be improved and made more sophisticated, but will there be anything radically new?

For one thing, there should be the growth of air freight, especially of bulky objects. In an age in which vast quantities of communication can be transferred electronically, the need for purely business air-flights by human beings will be reduced and freight may become as important a part of air transportation as it is now of rail transportation.

Then, too, particularly for smaller objects, there is underground freight. We can visualize a network of pipes (like those used for natural gas) underlying the nation. Through these, small automated containers can be directed along straightaways (by compressed air perhaps) and through switching stations. It would not be as fast as air-freight, of course, but it would not run the risk of crashes and it would be unaffected by weather.

In summary, then, if we look forward to a future in which mankind behaves rationally and avoids self-destruction, we can visualize a world that will be more complicated than the one we know today, but a world that will run better and, most of all, a world in which the individual will count for more, not less.

---

# 43

# The Coming Age of Age

Suppose that the world reaches the apocalyptic year of 2000 with society intact, with humanity secure, and with the future bright.

This is by no means a foregone conclusion. Though the year 2000 is not very far in the future, the various and assorted scourges of a rising population and declining resources, of growing pollution and the worsening quality of the environment, of increasing hunger and the diminution of available energy, all seem to make it appear that mankind is on a collision course with catastrophe and that the worst cannot be long delayed.

But suppose, nevertheless, that we make it. If we begin with that assumption, we can argue backward from that and determine what some of the characteristics of the world at the opening of the twenty-first century would then have to be.

We must, for instance, suppose that the population problem will, in 2000, have come near enough to a solution to allow the survival of our society.

At the present moment it seems that, barring unbearable catastrophe, the inertia of the current pattern of population dynamics will bring us to the year 2000 with 6 billion people on Earth. By then (we can hope) the growing weight of misery will have taught the world, in the harsh school of despair, that population must not rise further. Indeed, it must fall.

In that case, the year 2000 (the one we are imagining, the one in which mankind will be facing a bright future) will have to see a world in which the birth rate is everywhere falling, in which zero population growth is within reach, and in which negative population growth is the at least temporary goal being striven for.

What else will have to characterize the bright-future end of this century? Since there is no way in which we can avoid having used up much of the earth's oil reserves by 2000, we must conclude that alternative energy sources in quantity will be, at the very least, on the point of development.

There is a reasonable chance of this, actually. By 2000, geothermal energy, or direct solar energy, or nuclear-fusion energy, or even all three, may be pushing toward the service of humanity. This, in turn, implies that mankind will be supporting a still thriving and still advancing technological society, since the development of such sources cannot be carried through by Transcendental Meditation alone.

Other characteristics of a bright-future 2000 can also be reasoned out, but never mind; let us see what we already have, or can deduce, about this century's end (which will come when men and women in their early prime today will still be alive and vigorous) when we are facing a world of limited population and advancing science.

In a world of limited population, the birth rate must be lower than the death rate if, in the course of the twenty-first century, population is to decline from its intolerable high-point. This means that mankind will face a society in which the percentage of young people generally will be smaller than it has ever been.

This change in the makeup of the population will be accentuated if medical science succeeds in gaining a greater understanding of the aging process. Until now, through all the history of mankind, medical advance has only succeeded in making it possible for more people to grow old.

Medicine has won victories over those infectious diseases that kill before old age can sink its marks into the body; it has grown to understand vitamins and hormones so that metabolic disorders that do not involve the aging process can be corrected or ameliorated. But old age itself remains untouched. A man or woman who reaches eighty today is as old as one who did so in ancient times, and is as unlikely to live much longer. The maximum age remains today where it always was—not far beyond the century mark.

But if the biochemical and biophysical changes involved in aging become thoroughly understood, and if we learn how to delay or even reverse the process, we may, by 2000, face a future society in which men and women will routinely live to be over one hundred, perhaps far beyond one hundred. Nor will the stretched-out lifetime be one of additional decades of decrepitude and senility, but of vigorous middle age—continuing, perhaps, until the individual voluntarily decides to cease living.

Naturally, this will require ample quantities of food and good and thorough medical care; and in the world of 2000, which will be clearly overpopulated and only beginning to move into the pathways of sanity, there will be few pockets where the extended-lifetime potential can be massively met. It will be easy to see, though, that in the course of the twenty-first century, as the population goes down and energy grows more plentiful, the extended-lifetime pattern will spread itself over the surface of the earth.

This is another way of saying that the death rate will continue to fall, and that the birth rate will have to (*have* to) fall with it. Each year the age profile of mankind will show a smaller percentage of young and a larger percentage of aged. The twenty-first century, it would seem, will become, increasingly, an age of age.

Such a change in the makeup of the population is not a matter of choice, it is the price of survival; and it means that a drastic change in man's view of himself will become necessary.

Until now, until this very day, mankind has consisted primarily of young people, with those over forty making up a distinct minority of the whole. This was unavoidably true, considering mankind's life expectancy.

Through almost all of mankind's history, the life expectancy has varied somewhere between twenty and thirty-five (depending on time and place) and those who managed to survive all the rigors of life in the days before modern notions of medicine, hygiene, and social responsibility existed, and who reached the point where old age itself was the killer, were rare indeed. This had several consequences, including the following:

1. Old men, partly because they were rare, were valued. In the ages before widespread literacy (to say nothing of this modern day of electronic and computerized record-keeping), old men were an indispensable resource. They were the repository of the past, with memories that held the old traditions and the old ways. It is no wonder that it was to old men that society turned for advice, and that the brains and judgment of old men ran state and church. (The word "senator" is from the Latin word for "old" and "priest" is from the Greek word for "old.")

2. Old women were even rarer than old men, because they had to run the gantlet of child-bearing in addition to all the other ills that plagued mankind. Since old women were rarely seen and since their physical appearance was unusual, it is not surprising that they sometimes frightened ordinary men and women. The wrinkled face (old men's wrinkles were hidden by beards), the withered and stooped frame, the mumbled speech out of toothless jaws that, on closing, allowed the nose and chin to approach closely, is precisely the picture that grew up of the "witch"—a picture that still frightens children every Halloween.

3. Because old people of both sexes were so rare, they did not represent a great drain on society generally. Even when they were too old or sick to work, no social programs needed to be devised for them (as opposed to the great social programs of education that had to be devised for the young).

The numerical inconsequentiality of the aged might have continued forever if medical science had not interfered. In the 1860s, Louis Pasteur advanced the germ theory of disease, and from that time on the death rate began to drop. With continued medical advance, life expectancy rose until now the average child at birth, in many parts of the world, has a fifty-fifty chance (or even slightly better) of living to be over seventy.

This means that in the past century the number of old people has rapidly increased both in absolute numbers and in the percentage of the population — particularly in the industrialized nations. What's more, by drastically reducing the dangers involved in child-bearing, a woman's life expectancy has become greater than that of a man's, so that there are today considerably greater numbers of old women than of old men.

The same scientific expansion that extended life expectancy, also provided copious mechanical methods for keeping records with far greater accuracy and efficiency than is possible for merely an old man's memory. Old people, no longer valued (or feared) because of rarity, also lost their function.

Again, the advance of industrialization lessened the need for manpower. Then, too, the urbanization of mankind, and the manner in which increased mobility made it possible for the units of the family to scatter, reduced the kind of chores on the farm and in the family that had always been available to the old. What's more, the rapid accumulation of knowledge, the rapid advances in techniques, the rapid changes in the state of the various arts, meant that the experience of age was a liability rather than an advantage. The wisdom of age fell victim to the up-to-dateness of the downy-cheeked recent graduate.

Industrial societies have the chronic problem of keeping unemployment down. One way of fighting it is to introduce the concept of enforced retirement at some age, such as 65, and thus, artificially, make room for younger men. The value of doing so is that a young man without a job is unemployed and must be counted as such, while an old man without a job is retired and does not enter the unemployment statistics. (As a reward, he is bribed, minimally, with Social Security, to keep him from embarrassing society by starving.)

The result is that, through a complex chain of circumstance, we face today a society containing an unprecedented number of old men and women, whom, out of long tradition, we make no effort to treat humanely and creatively. Instead, we deprive them of the work that might continue to give their life meaning, and condemn them to the miserable task of doing nothing but wait for death.

Shall we continue to do this when, if we are to survive as an organized society past the end of this century, we must raise the percentage of aged to a hitherto unheard-of-level? Even if the total number of old men and women declines along with the declining population, their percentage of the population will go up substantially.

Surely society will have to alter its youth-centered attitude. It will have to become age-blind. (It may be that we will even have to drop those terms that, sanctified by the ages, serve to carry the connotations of centuries of prejudice—"aged," "elderly," "old." We may have to speak of the "post-youth.")

Consider, for instance, the matter of education. Throughout history, it has been assumed that education, in any formal sense, is for the young only. Education usually stops when the young man or woman has learned enough to support himself somehow or has mastered the rudiments of those clichés that will make it possible for him to carry on polite conversation.

Individuals, once past their youth, if they continue to wish to learn, must generally do so on their own time and by their own efforts. If they attempt the task in any organized fashion, they are competing with the young and do so, in our youth-centered society, at a psychological disadvantage.

As a result, those with a specialized education find that, in a time of rapid technological change, they grow rapidly obsolete. Even those whose contribution to society does not suffer adulteration with time are slow in adapting themselves to changing circumstances. With education confined to youth only, they retain the attitudes, beliefs, and clichés suitable to their youthful, and only, years of education, so that they become the Archie Bunkers whose outmoded ways of thought amuse young people and convince them that there is something inherently stupid about old people.

It is a vicious cycle in which we deprive the post-youth of socially approved opportunities to learn, allow the faculty of learning to atrophy, conclude that "old dogs cannot learn new tricks," and then use the conclusion to justify a continuation of the procedure that makes the conclusion inevitable.

How can this continue in a society in which the post-youth will be predominant? Universities will then have to be open to all on an equal basis, regardless of age. Nor can the older people be stuffed into the ghetto of adult-education programs in a way that cannot help but make them seem faintly amusing to the world in general. People of any age will have to be allowed precisely the same opportunities to profit by any kind of education under precisely the same conditions.

But is it, perhaps, unfair to expect the old to compete with the young in class?

The question at once assumes that the old are less capable of learning than the young are, so that it would be a kindness to the post-youth to segregate them. How do we know that this would be so in a society in which the older individuals take it for granted all their lives that the opportunity to learn will never be denied them, and that it will always be assumed that they can learn if they wish to do so? How much of the apparent learning incapacity of older people today is due to the fact that they themselves have always assumed they would be increasingly incapable of learning as they grew older?

Then, too, in what way is the educative process a competition? We might assume that each person will be educated, primarily, in order to live his or her own life more fully and to make his or her unique contribution to society more effectively. Since the human mind is sufficiently complex and versatile to make no two human beings completely alike in talents, abilities, attitudes, and desires, where is the competition?

Remember, too, that if we are assuming a bright-future world in which science and technology are advancing, we must also assume that computerization and automation will advance. We will therefore be facing a century of declining necessity for finding hands to do the dull, repetitive, and meaningless labor (both physical and mental) that through all of man's history has stultified his soul.

The importance of education for worthless jobs that machines can do better than men, and that are therefore beneath human dignity, will decline. Education will become more and more a matter of teaching creativity; or, better, supplying the environment, atmosphere, and materials that will make it possible for each person to maintain creativity within himself.

It is a mistake to think that only a few, a very few, fortunate individuals are creative. Perhaps only a few, a very few, are transcendentally so, but all people of even nearly normal intelligence  show remarkable creativity in their early years. The ability to learn to speak, to read and write, to do even simple arithmetic, is a demonstration of enormous creativity. If this creativity falls off with years, it is more because the school system, as now organized, is designed to stamp it out than for any other reason.

And, in a world that is thoroughly computerized and automated, there will be enough who will want to do the work of computer programming, scientific research, educational innovation, medicine, law, government, art, music, literature, and so on, for no other reason than that they want to.

And there will also be others who will want to do other things, such as playing cards, or bird-watching, or building model skyscrapers out of toothpicks, for no other reason than that *they* want to.

But what's the difference? All will be doing what they want to and all will be contributing to the joy of the world, generally.

And with the opportunity of changing careers, of trying different "sub-lifetimes" in one extended one, and knowing that radical changes can be made at any time and that there will be time for a full preliminary education in preparation for each change, men and women can retain the qualities of creativity and eager innovativeness all through their lifetimes. There may be no reason, we will find, to suppose that because a society has few young, it will be stodgy and stagnant.

If this is to come about, however, considering that we will be entering the age of post-youth society in the lifetime of individuals alive today (assuming our society survives at all), it is time to begin revising our attitude toward age *now*.

Yet perhaps there is more to fear, in an age of age, than possible social stagnation. Where people live extended lifetimes, the gene patterns do also.

The rate of evolutionary change depends on the frequency with which new gene-patterns are produced and the rate at which the processes of natural selection can choose among them. With few children being born and with a slow turnover of generations, might it not be that, regardless of how innovative old people can be, our species will nevertheless become *evolutionarily* backward; that it will not change to fit the changing environment and altering conditions; that it will be embarking on a course that for other species through Earth's long biological history has inevitably led either to extinction or to a nonprogressive, changeless lingering in some static evolutionary niche.

But man is *not* other species. Thanks to medical research, advanced biological engineering techniques may be developed. The future society of man may then demonstrate a new kind of evolution: not that brought about by numerous deaths and births and the random directives of natural selection, but that brought about by intelligent choice and design, moving steadily and humanely in the direction we think it desirable to go.

This will be something completely new under the evolutionary sun; but, if we are to survive at all, *everything* will be completely new, and we will have to learn to deal with it all.

It will be a formidable task—but what an exciting one!

---------------------------------------- **44** ----------------------------------------

# The Decade of Decision

We are now in the 1980s—the decade of decision.

Every decade is a decade of decision in one way or another, but this one is different because it involves the whole world and because it involves the life and death of civilization.

It was possible in past periods of history for a particular civilization to decline while others remained flourishing. When western Europe was in the "Dark Age" between A.D. 500 and 1000, southeastern Europe, northern Africa, and southern Asia were doing very well.

And it was possible for the civilization that did decline to recover and move even higher than before, as western Europe began to do after A.D. 1000.

This may not be possible now. Thanks to the steady advance of science and technology, the world has become so small that it is now a single economic unit with the welfare of each advanced portion of it dependent on all the rest.

Thanks to the steady advance of science and technology, the stakes are now so high, the requirement for material resources and energy so intense, the numbers of people and the intricacy of the environment in which they live so great, that any stumble could mean so huge a smash as to make recovery impossible.

And, thanks to the steady advance of science and technology, the means for destruction in the hands of human beings is so great that the world could be ultimately devastated for thousands of years to come if we were to try to fight a war with all the power at our disposal.

But, then, thanks also to the steady advance of science and technology, there are possible solutions ahead if we care to take advantage of them.

Let us first list the major crises we will face in the 1980s.

First and most immediately crucial is the matter of energy. Our civilization runs on energy and always has. The history of humanity is that of the slow, but steady, increase in our ability to make use of greater and greater sources of energy, until finally, in the eighteenth century, we learned to make use of fossil fuels and to use them to run steam engines. With that began the modern age of industrialization.

The Industrial Revolution was powered by burning coal throughout the nineteenth century. That was good enough for the steam engine but the internal-combustion engine required liquid fuel, and in the twentieth century, with automobiles and airplanes leading the way, the world began to shift to oil. This accelerated and  after World War II oil became the chief power source the world over. It was the most convenient and, for a while, the cheapest energy source the world had ever seen, and for thirty years the planet boomed.

But there is only so much oil in the ground and, at the present rate of use, it will all be gone in between thirty and fifty years.

At the *present* rate of use. If the rate of use increases, the rate of production cannot for very long keep pace. During the 1980s, the two lines could cross and production may fall below what people would like to be able to consume.

In fact, the 1970s were shaken by the advance tremors of the earthquake. The price of oil has risen steadily and rapidly since 1973. To be sure, this seems to be the result of the voluntary action of the oil-producing (OPEC) nations, but they are able to raise the prices when they do only because they know their customers *must* have the oil and that it will be in increasingly short supply.

Even if somehow the OPEC nations could be persuaded to sell their oil cheaply, or give it away for free, that wouldn't help. It would only mean that the oil would be used to a greater extent than ever and would run out sooner, and then would come the real disaster. In fact, it might be argued that OPEC is doing the world a favor by raising the price of oil. It encourages the world to use oil more sparingly so that it will last longer, and it also encourages the world to seek alternative sources of energy.

That is the first crucial life-and-death decision that the world must make in the 1980s.

Either the world will continue to rely on oil, burning it as quickly as it can, at any price, until there just isn't enough to be had so that the whole world breaks up into scavenging bands, each trying to survive at the expense of the others, and civilization is destroyed — *or else* the world must make the deliberate decision to spare oil, lower waste, and needless luxury use, and develop alternative sources of energy.

That *or else* decision must be reached before the end of the 1980s, or it could be too late to avoid catastrophe.

It is not likely that people will be able to concentrate on the great issue of energy and survival if their attention is distracted.

Through all of history, there has been a tendency for nations to try to solve their problems through war. If the food supply is low, rob your neighbor; if the restlessness of the nobility threatens the stability of a nation, send them off on a crusade; if a well-trained army has nothing to do and is getting rusty, start a career of conquest.

Long-established habits are hard to break. Even if a nation realizes that it is no longer rational to make war, it will still go through the motions of preparing for one. After all, each nation asks itself, who knows if a neighbor nation might not be preparing for conquest? Besides almost every nation has some grievance against its neighbor, so that one seeks "justice" and the other fears "revenge."

The 1970s saw a number of wars in the Middle East, in the horn of Africa, in Southeast Asia. Even where there is no active war, terrorist activities are to be found in Northern Ireland, in Spain, in Italy, in Germany, and so on.

And every year, the world, as a whole, spends $500 billion on armaments.

Even small wars are destructive and they always threaten the possibility that the superpowers may be drawn in, and that a confrontation of ordinary missiles and explosives may turn into a war of nuclear weapons. Even if there were absolutely no wars, large or small, and no terrorism, as long as the nations prepare for war at the present rate, they waste energy in enormous quantities and hasten the day when energy-lack may topple civilization.

Besides, as long as every small group of people concentrates on its own short-range goals, its own "justice," its own "revenge," its own "national security," there will be no attention paid to the overall problems of survival that face civilization and humanity. And when civilization crumbles and billions die and humanity is reduced to broken bits of scavenging bands of tattered barbarians, where will all that justice and revenge and national security be?

So that is the second crucial life-and-death decision the world must make in the 1980s.

Either it will continue to behave as though the world were still a nineteenth-century conglomeration of nations, warring and preparing for war, each trying to improve its standing at the expense of the others, each trying to dominate its region, its continent, or the whole planet—*or else* the nations must make the deliberate decision to stand together and, understanding that the crisis in energy could put an end to all civilization everywhere, agree to a planetary attack on the crisis, leaving all minor problems to decision by arbitration, compromise, goodwill, and patience.

That *or else* decision must be reached before the end of the 1980s, or it is likely to be too late to help.

Even new sources of energy, no matter how copious and cheap, and even a cooperating group of nations, no matter how sincerely friendly, are not likely to solve the world's problems if the population of the earth continues to rise. If we continue to add somewhere between 1 and 2 percent to the population each year, the present 4,000 million will become 6,000 million by the year 2000. In the 1970s we added about 700 million people to the population of the world, and in the 1980s we may add 800 million.

We are not feeding the population of the world adequately now; we are sure to feed them less adequately if we add the equivalent of another India to the world in the 1980s. It will be all the worse as the energy supply becomes tighter and tighter, for one of the reasons we can manage to feed the world as well as we do right now is because, in some parts of the world, farming machinery, irrigation pumps, fertilizer, and pesticides are used. All of this requires much energy, and as the availability of energy declines so will the food supply.

This, then, is the third crucial life-and-death decision the world must make in the 1980s.

Either it will continue to increase its population so that there will be more and more starvation and an increasingly desperate need for food that will cause humanity to place more and more pressure on the environment (further reducing its food-growing capacity) until the nations of the world go to war with each other or break up into scavenging bands with each scrabbling for food that isn't there—*or else* humanity must make the deliberate decision to control population by limiting the birth rate everywhere.

That *or else* decision must be reached before the end of the 1980s, or it is likely to be too late to help.

It is easy to say *or else,* but how exactly can we implement that *or else*?

In the case of energy, the widespread use of oil, its cheapness and convenience, has allowed us to atrophy the use of other energy sources, but we can go back to them. We can build more and safer fission power plants, develop hydroelectric power, increase the use of coal or oil shale, grow plants especially designed to supply fuel, build windmills, make use of tides, of geothermal energy, and so on. All this would be difficult and would involve a good deal of effort and capital expenditure, but every bit we do in this direction will tend to conserve oil, stretch out the supply, and give us more time to further develop the alternative sources.

Meanwhile, we can use the extra time to obtain energy in completely new ways that may be even more convenient and cheaper than oil ever was.

One possible route is the development of nuclear fusion. We don't have it yet, even in the laboratory, though scientists have been working on it for thirty years, but we may find the answer in the 1980s—even though it will then take some decades to translate it into large, practical power stations.

Another route is the development of solar energy. This can be done, to begin with, in small-scale fashion in individual buildings, and that will spare oil. Eventually, solar energy can be used on a larger scale by coating large desert areas of the earth with photoelectric cells.

On a larger scale still, we can build solar power stations in space. We can imagine square miles of photoelectric cells exposed to radiation in space, in orbits that will place them in the earth's shadow for no more than 2 percent of the time and where there will never be any atmospheric interference at all. Numbers of such solar power stations can be built, and they may be an answer to the earth's energy problems for billions of years—if we last that long.

It's possible! The United States and the Soviet Union have shown that human beings could remain in space for three to six months at a time without bad effects. What's more, the United States has developed a "shuttle" spacecraft that can lift masses into space and return for more. The reusability of such a craft will greatly lessen the expense of spaceflight.

It is therefore possible now to make up our minds to begin a project of putting space to use—of collecting energy in space and beaming it down to the earth; of building space settlements to house men and women in small independent worlds of their own and leaving the building of further energy stations and other structures to them; of building a mining station on the moon to supply the metals, concrete, soil, glass, and oxygen for such settlements and such structures; of building observatories and laboratories and factories in space—in general of expanding the range of humanity from the world in which it has been confined through all its history to a much larger realm.

All this cannot be accomplished in the 1980s, but a beginning *must* be made. If, in the 1980s, we do not decide to extend our reach into space, then thereafter it may well be too late. We will not have time to overtake catastrophe.

Can it be done if humanity insists on spending its time, its effort, its resources, its emotions on nineteenth-century national problems?

No, it can't; but then the true reach into space is beyond the capacity of any one nation in any case. It is a global project, and it should be encouraged as such and used to stimulate world cooperation.

If the peoples of the world recognize the great crisis of survival that faces them and further recognize the importance of building a space-oriented economy and society—of getting energy from space for *all* nations, for instance—then they may well be anxious to work together for it. Each nation, in fact, may fear being left out, may fear being unable to lay claim to its fair share of the benefits to be expected from space.

The United States and the Soviet Union, with the major space experience and the major capacities in this direction, should encourage cooperation between themselves and participation of all other nations, in however minor a capacity.

Around this, there may be built the framework of a world combination that will gradually develop into a federal government for dealing with world problems. Individual nations can continue to deal with their own special problems in their own special way. Indeed, if the world combines on the major problems, there will be no harm if the various nations break up into smaller units. It won't matter how many "nations" there are, how many newly independent Quebecs, Brittanies, Scotlands—if there is only one world government.

A world government cannot be accomplished in the 1980s, but a beginning *must* be made. If, in the 1980s, the nations of the world do not decide that cooperation for survival is an absolute requirement, then thereafter it may well be too late. We will not have time to overtake catastrophe.

Can all this be done, however, if population continues to increase?

No, it can't; but there are signs of a growing understanding of the problem. In more and more regions of the world, the birth rate is dropping. Sometimes it only requires the release of social pressure in favor of children. If the feeling grows that there is no moral disgrace in having few children or even no children, then the birth rate might drop at once. Allow women greater freedom to participate in the great work of the world, and the birth rate will drop. Adjust the tax pattern appropriately, and the birth rate will drop.

Since the birth rate has been dropping since the late 1970s, we can hope that the pattern will continue in the 1980s—everywhere.

We will not bring a halt to population increase in the 1980s, because it is hard to move great masses, but a beginning *must* be made. If, in the 1980s, the world does not decide that a great priority must be given the lowering of the birth rate by every humane measure possible, it may be too late thereafter. We will not have time to overtake catastrophe.

And, of course, if we place limits on population, the time will come when we can move more and more people out into space settlements—and then, out in space at least, population can grow again for a time.

Then, too, on the side of survival is the continuing development of science and technology—methods for increasing food output, for safer contraception, for more efficient energy use, for better communications. Most of all we have the developing computer, growing ever more complex and versatile and capable of answering more and more difficult questions.

Problems that seem insoluble now may become soluble with properly programmed properly complex computers.

Well, then, which decision will the world make in the 1980s? How can one predict? It depends on the balance between the rational and the irrational, on the battle between the need to take a new path and the fear of leaving the old one, on whether there will be the imagination and vision that will leave the old quarrels behind, or the ingrained hatred that makes one content to die provided one has one's revenge.

I hope we will be rational, that we will change to the new, that we will have the vision to leave the old behind—but I cannot be sure we will.

At any rate, by the end of the 1980s, I feel that we will know what the decision has been.

# 45

# Do You Want to Be Cloned?

The matter of cloning has been much in the news, and there has been considerable controversy over it.

Cloning is not new, however. It is as old as life, and man has known about it as long as he has been cultivating plants. A clone is an organism that has been produced from another organism without the intervention of sex.

A bacterium can divide over and over to produce any number of additional bacteria. All those bacteria are clones of the first. A twig of a tree can be planted, and may develop roots and branches and become a new tree. It is a clone. (In fact, the word "clone" is Greek for "twig.") A starfish can be torn into several pieces, and those may be thrown back into the water. Each piece will then grow into a complete starfish and all will be clones of the original.

The more complex animals do not form clones. They only reproduce sexually. Females produce egg cells and males produce sperm cells; each cell containing a half-set of chromosomes. A sperm and egg combine to form a fertilized egg-cell that contains a whole set of chromosomes, half from the female and half from the male. The fertilized egg-cell then divides and redivides and eventually forms a new organism. Such a new organism, born of sexual combination, has two parents and is *not* a clone.

Sometimes, though, a fertilized egg divides in two and the two new cells come apart. Each of the separated cells then goes on to divide and redivide and form a whole organism. This can happen in the case of human beings, too, and identical twins result—same sex, same appearance, same chromosomes. Each of the identical twins is the clone of the other.

251

As a fertilized egg divides and redivides, individual cells lose the ability to give rise to a complete organism if separated. Each cell that forms from a fertilized egg retains copies of the original chromosomes, but those begin to be modified by outside influences. Some parts of the chromosomes are blocked, others are stimulated. In the end we have skin cells, liver cells, heart cells, kidney cells, lung cells, and so on, all with the same chromosomes, but each having them differently specialized. All are so specialized they can't divide to form a new individual. Some are so specialized they can't divide at all.

In each body cell, the chromosomes are contained in a small portion called the nucleus, which is seperated from the rest of the cell by a membrane. Suppose you separate this nucleus from the body cell and transfer it to an egg cell whose own nucleus has been removed. Under the influence of the material in the egg cell, the genes in the nucleus of the body cell are unblocked. Now the egg cell can divide and redivide to form an organism — one in which the chromosomes come from the person (male or female) who donated the cell nucleus. We have a clone of the donor, with the same chromosomes, the same sex, the same appearance.

Can it be done?

It *has* been done in some animals. In the early 1960s, clones were produced from the cells of tadpoles and, in 1975, from the skin cells of an adult frog.

Can it be done with animals that are closer to the human being than a frog is? With mammals, such as rats or rabbits? With human beings?

So far it hasn't been done. Mammals have egg cells that are considerably smaller than frog eggs and that are more delicate and more easily damaged. What's more, a frog egg can be put back in the water and allowed to develop there, but a human egg must be inserted into a female and allowed to develop in her womb.

In March 1978, David Rorvik published his book *In His Image,* which purports to tell the story of the cloning of a man, but people in the field simply did not take it seriously. The state of the art, they all agree, is not up to such a feat, and the book has been declared a hoax.

Yet biologists are sure that some day they *will* be able to clone mammals, and even human beings. In that case, there are questions we might ask? Is cloning good or evil? Right or wrong? Useful or dangerous?

In coming to a decision we have to know what cloning can and can't do. Some people have the idea that clones offer a gateway to personal immortality, for instance, and want it for that reason.

Not so! A clone is not *you.* A clone would only be an identical twin brother or sister, born late in your life, and it would have his or her own distinct personality and identity. Suppose there are identical twin brothers (or sisters) born in the usual way. If one of them dies, the dead one does not live on in the twin even though both have the same chromosomes. The dead one is *dead*—and so it will be if you are cloned.

Some people think that cloning is dangerous because it will enable aggressive governments to produce hordes of docile people of subnormal intelligence to serve as laborers or soldiers. This is a useless fear. No government ever found it difficult to collect laborers or soldiers even without cloning, and it is far cheaper to produce them in the normal way than by cloning.

Remember that a clone is not only hard to produce but that it takes just as long to produce a clone as an ordinary human being. A clone must start as an egg, be nurtured for nine months in some woman's body, and then take the usual eighteen years to be old enough to vote or fight.

Well, then, how about using clones to reproduce genius? We can always use additional Einsteins, Picassos, Beethovens, and Tolstoys. If such great people have children in the ordinary way, their chromosomes are mixed with those of their mates, and the combination may not represent quite the genius of the one parent. If we clone a genius on the other hand, we have new individuals with the precise chromosomes of that genius.

Will we then have fifty geniuses who can produce the great works of art, literature, or science of the original? Very likely not, since human beings are not the product of their chromosomes alone. In cloning, a nucleus must be put into an egg cell and then into a womb, and the matter in the egg cell and the nature of the womb will have its influence.

Then, too, the clones are born anywhere between thirty and fifty years after the person who is being cloned was born. Everything has changed in the interval. The clone will not be presented with the same opportunities and the same obstacles that the original was, or be part of the same society. Each will go its own way, and not a single one may duplicate the genius of the original.

In that case, of what value is cloning?

Well, biologists do many experiments on mice, rats, guinea pigs, monkeys, and other animals in order to gain information that might be applicable to human beings. The experiments give us important information on nutrition, on medicine, on behavior.

One of the possible confusions about such experiments is that different animals of the same species have somewhat different chromosomes and may respond differently because of that. If different rats, for instance, are subjected to different conditions and react differently, is the difference in reaction caused by the difference in conditions or by differences in the chromosomes? We can't be sure.

If, however, we were to clone a rat over and over, we might have fifty rats with identical chromosomes; and, if we experiment with them, we know that the difference in reaction would have to be caused by the difference in the conditions.

Again, suppose we develop methods for removing or altering a particular piece of a chromosome and want to compare two rats so as to see what that one little change will do. If the rats have different chromosomes to begin with, introducing one little change might not yield clear-cut results. If we make use of clones in which all the rats have identical chromosomes, and then introduce a small change in one rat, a different small change in another rat, and so on, our knowledge of just how chromosomes do their job will surely leap ahead rapidly.

Then, too, there are many species in the world that are now endangered, whose numbers have grown so low we are not sure they will survive much longer. Cloning may be a way out for those we most value — a way of producing more of them, using, if necessary, females of an allied, more numerous species as host-mothers.

It has even been suggested that, when a frozen mammoth is discovered in the icy soil of Siberia, some cells might still be alive enough to be cloned. An elephant egg-cell might be used to house the mammoth nucleus, and an elephant mother might nurture the clone in her womb so that, after two years, the elephant might give birth to the first living mammoth the world will have seen in at least ten thousand years.

Human clones might be treated in the same way. It might be useful to experiment with human clones in order to gain theoretical knowledge concerning nutrition, medicine, and psychology. Experiments with human clones, however, involve such serious ethical questions that it is doubtful whether they can be carried through — or should be carried through.

There is one application of human clones that may be hard to resist, however.

By working with clones of the lower animals, we might learn how to develop mammalian embryos in laboratory equipment without the use of a woman's womb. We might learn how it is that an embryo's cells develop and how they specialize into different organs. We might learn how to treat embryos in such a way as to alter the normal development and cause them to give rise to a heart in particular, with everything else vestigial; or a lung; or a kidney; or a leg. Perhaps, if the developing embryo is forced in that direction alone, a full-sized adult organ might be developed in, let us say, a year.

If, then, the time were to come when an adult found he had a limping heart, a fading pancreas, or a leg that had been lost in an accident, some of his skin cells might be cloned and made to grow a replacement organ.

An ordinary organ transplant from an ordinary donor is often rejected by the very body that can't live without it, because the transplant has a chromosomal makeup different from that of the body. An organ replacement built up out of the body's own cells would have the same chromosomal makeup as the body, and the body would not reject what is, after all, its own.

A clone may not make you immortal, but it could, in this way, at least extend your life by giving you the equivalent of a spare heart, or kidney, or whatever.

---46---

# The Hotel of the Future

In any discussion of the future, we must remember that there are an infinite number of possibilities and that humanity has the ability to choose among them.

Let us therefore suppose that we have chosen *not* to bring down doomsday upon ourselves. We will have no nuclear war; no overpopulated, over-polluted starvation. Let us suppose a working civilization of continuing technological advance.

In that case, we must remember that we are living in a society that is, right now, undergoing an enormous revolution, one that is not entirely clear to us only because we are immersed in it and lack the perspective that distance in time will eventually bring.

I am referring to the matter of computerization and automation.

It is surely inevitable that there will be ever wider use for plastic devices (like credit cards) keyed more and more to some system for personal identification and, second, to a national (or even global) information network.

We are rapidly approaching the time when a large and growing percentage of American homes will have a personal computer-terminal, and one of the many functions of such a terminal would be the making of hotel reservations.

A credit card would be inserted into an appropriate slot and the computer keyed for the making of a hotel reservation. The computer may, for the sake of security, require a special identifying number — or may make use of some more subtle manner of establishing your identity, such as a voice-print or a thumbprint.

Once that is done, you will feed the necessary information into the computer. The name of the hotel and city, the date and time of arrival and departure, the number of people in your party, and any special requirements you may have.

A message will flash back at once, accepting your request, and indicating the room number in the hotel of your choice over the period you have indicated. If you confirm that, the data is entered in the hotel books and you have the room. If the hotel is full over the period indicated, comparable accommodations in another hotel in the vicinity can be offered, which you may or may not accept.

At any time prior to the moment of leaving, you can insert the card, call up the information on the screen, and modify it or even cancel it if necessary. Or you can check to see whether the hotel has had to modify or cancel the reservation.

At the hotel, computerization will continue. The long, slow lines at the registration counter will no longer be there. Instead, there will be a much shorter and faster-moving line at the hotel's registration terminal. When you step up to it, you will insert your card, which will mean that your name and other pertinent data will be instantly entered into the hotel books.

You will, in return, receive your card-key (another plastic rectangle, or two of them, if you have requested a separate card-key for your spouse in your reservation). In some hotels, it my be necessary to speak into the registration terminal to get your card-key, in order to have your voice-print on record.

When you arrive at your room, you will insert the card-key in the appropriate slot and speak briefly. You will then remove the card-key and the door will slide open. (Undoubtedly any number of people will choose to say "Open sesame," especially if in a jovial mood. Others, more serious, are apt to repeat the set-phrase that they recorded in getting the card-key, saying, for instance, "My name is John Doe and my room number is 000," in order to avoid the inconvenience of a transitory misidentification.)

From the inside, the door will open merely by turning the doorknob. (There will be no doorknob on the outside.) There will, of course, be a chain or bolt on the inside of the door for additional security.

There will be a similar saving of time on leaving the hotel, since once again there will be no long, slow lines at the cashier's desk.

After all, we will be living, increasingly, in a cashless society in which the use of an elaborate credit-card system will at once transfer money from one account into another.

Your card-key will be inserted into the slot provided for the activation of the cashier terminal. Your credit card will then be inserted into the terminal slot and, at once, the screen will flash the details of your bill. (Naturally, the screen will be in a small booth for the sake of privacy.)

If there are no questions concerning the bill, you will indicate that by pressing the appropriate contact; and cash will immediately be subtracted from your assets and added to those of the hotel. (That portion of the payment that represents tax will, of course, be added to the assets of the government—and will be taken into account at income-tax time.)

Of course there is the possibility of a hitch. You may indeed question one or more items, or the terminal may find you lack the assets to pay the bill. (The wise traveler will constantly check the state of his credit-balance—something he can do at home at his own terminal, or in any public terminal, of which there will be many, for a nominal sum. Still, even the best of us can forget or misread and end up embarrassed at a terminal that flashes: INSUFFICIENT ASSETS. It seems unlikely that anyone would attempt outright fraud, since that could never get past the terminal without extensive expertise.)

In case of a hitch, it may be necessary to open negotiations with a human representative of the hotel, but clearly this will not often be required.

Then, too, hotels will be increasingly robotized.

This is not surprising. Industrial robots have become a major force in our economy and their use is expanding rapidly. What's more, robots are becoming more versatile and "intelligent" with each year, and the larger robot-manufacturing concerns are working very hard to develop home robots in order to enter what is, potentially an enormous market.

It is hard to tell just what these robots will look like, but I doubt that they will be humanoid in form, certainly not at first. They would be expected to be simple and functional.

For instance, there would be a robot bellboy—essentially, a self-propelled baggage container. A group of them might be waiting in the lobby. You will insert your room-key in an appropriate slot, and a panel will slide open. You will place your baggage inside (using more than one compartment, if necessary.) It will close again, and you will go on to your room.

Once the robot bellboy is filled (or even if some of its compartments are not filled, if a given time has elapsed without another person signaling for its services) it will move on to its special elevator bank.

It will need no directions, since the use of your room-key will have informed it of the destination of your baggage. It will be at your room not long after you are. There it will signal for entrance, and you can remove your baggage.

It is easy to imagine improvements. If such a robotic device is equipped with a sense of vision, and with manipulative appendages, it will be able to see your baggage and load itself. Once in your room, it can hang clothes-bags in the closet and place suitcases on the special racks provided. It might well be equipped to inform you of the amenities of the room and methods for making use of them.

Other services in the room can similarly be robotized. A cleaning robot will appear and, if the room is unoccupied, it will open the door by appropriate signal and enter it to clean it up.

It will have vacuum-cleaning attachments, and it will be able to pick up objects and decide whether to dispose of them, or place them in special containers for your attention. (The decision will be heavily weighted in favor of the latter; for only the most obvious of trash can be disposed of, since it is better to save something useless for you than to throw out something valuable that you have inadvertently dropped.) It will empty and clean out ashtrays and wastebaskets, wash the bathroom floor and fixtures, collect and replace towels and washcloths. It would even make the beds neatly.

You may wish the cleaning robot to do its work under instruction when you are in the room (or have it work under special instruction even when you are not). That will be possible, for there will be a computerized control-panel in your room on which you can indicate the time-range within which you want the room cleaned, and what, if anything, you wish done or not done.

The computer in your room will be, in fact, your full companion; silent, attentive, and unforgetting.

It is the computer you will alert for a wake-up call at some particular time; the computer with which you can leave room-service requests or order valet-service, specifying time and details; the computer on which you can record any television programs you will want to watch (time, channel), any telephone calls you will want to make, specifying the number, and any messages you will want to leave for those who call you when you are not in your room.

Most of all, it will answer your questions about the area; the entertainment available, plays or concerts, times and places of everything, shopping opportunities and locations and methods or routes of travel. It will arrange for a taxi to be waiting for you at the hotel entrance at a specified time or will tell you if that cannot be done and when it can be.

Naturally, if you have requests that require delivery, that will be done robotically. Your servers and valets will be robots that will bring your dinner, set up the table and spread it, and eventually return for the leavings. Or they can arrive to accept shoes, linen, suits, or whatever, and eventually return them polished, washed, pressed, as the case may be—always supposing them to have been properly instructed.

What if the computers and robots break down? Then they must be repaired.

There is always a chance of breakdown and, realistically, we must live with it. Right now, there is always a chance of a citywide, or regionwide, electrical blackout, but no one seriously suggests that we avoid such a thing by abandoning the use of electricity.

Naturally, all services will represent items on your eventual bill, but you will be spared the necessity of tipping and, for that matter, the invasion of privacy that human beings always bring with them, however quiet and unobtrusive they may be.

In the ideal hotel of the future, you will never have to encounter any human beings except those you want to encounter.

Is this a dehumanizing prospect? Not really. We have been moving in this direction for a long time. When we take a push-button elevator, we don't feel uneasy over the absence of an operator. When we make a phone call, we don't ordinarily long for a messenger to send instead. When we turn on the television set, we don't sigh for the days when we might have been a king and could call in the jester.

We don't really want all the attendants we currently cannot do without. We try not to notice them when they appear, and are relieved when they go.

What about the problem of unemployment? What about all the bellboys, chambermaids, and hotel employees generally that will be thrown out of work?

In the short run, they will represent a problem that society will have to deal with. In the long run, the whole matter will be a social service. The kind of jobs that robots can replace are menial, repetitive, and stultifying ones. It is hard to get anyone willing to do them except out of the sheer necessity of earning money, so that they are usually done without enthusiasm or care.

In a robotized society, there will be new jobs involving the design, manufacture, repair, and maintenance of robots that will require more skill and provide more interest and excitement than any job a robot can take over.

A second revolution that will affect the role of hotels in the future rests in electronic communications.

Communications satellites already exist, and they will grow more numerous, more versatile, and more capable. Optical fibers are already replacing metal wires so that messages can be carried by laser beams of visible or ultraviolet light instead of by radio waves. This means that the number of wavelengths available for messages will increase by millions and tens of millions over the number available now, thanks to the fact that light waves are millions of times shorter than radio waves.

Closed-circuit television will become a much simpler matter. Five people on five different continents could hold a conference not only with sound, but with three-dimensional sight, through holography.

There will be less and less need to transport mass when all we need is the information the mass controls. No need to transport the actual letter, when we can wire the message and produce a facsimile at the point of destination. No need to send a person when his image can deliver the message, or when he can, from his home, observe, direct, and control machinery in an office or factory.

In short, there will be less and less *forced* travel, with people moving unwillingly from place to place for busines reasons, to meet someone, or to do or find out something. A larger and larger amount of total travel-time will be occupied by people who are traveling for fun—to see sights and do things they cannot see and do at home, or to meet friends and hug something more than an image.

Travel will become almost entirely tourism, in other words; and hotels everywhere will be organized with this in mind.

Hotels will actively organize and supervise tours and arrange for entertainment. They will present lectures on the city or the region in which they are located, educating those who have come there so that they may play the role of tourist more intelligently, and gain more from it.

In fact, there may be tourists who will want as much of the new place brought to them as is possible—the food, the entertainment, the nightclubs, the shops, the atmosphere. Even those things that can't be moved—the Taj Mahal, the Grand Canyon—may be preferred in holographic three-dimensional image.

For that reason, hotels may become more than ever self-contained worlds that, ideally, can offer their guests a full measure of the *sense* of being in New York, or Paris, or Bucharest, or Rio, or Naples, or Calcutta, without their ever having to leave the hotel.

What's more, when hotels truly appreciate their function as thoroughly computerized and robotized tourist-paradises, then they will also understand the advantages to certain novelties in location.

What of an enormous hotel that is located underground? There would be such a sense of privacy and enclosure, of separation from the workaday world. Many people might find that appealing—certainly for a limited vacation period.

An underground location would offer an environment totally free of weather, too. The temperature would be equable and cool, without the vagaries of air-conditioning. It would never rain or snow, and the wind would never blow fiercely. (Well, there might be some leaks if there were heavy rains overhead.)

Or how about a hotel on the Antarctic rim? Modern technology could keep such a hotel supplied with the necessary comforts for its guests, and again there would be a sense of privacy and separation. And, for those who enjoy such things, think of the skiing.

Or a hotel on the continental shelf, not too far off-shore. Such a hotel could be approached by launch and entered by way of a large "snorkel" thrust above sea-level under non-storm conditions; or it might be entered most directly, by submarine.

In that case, think of the water sports—fishing, swimming, water-skiing, scuba-diving.

And, of course, there will be hotel people who will be thinking of the final conquest, and who will be hiring engineers and architects to design the ultimate hotel, one that is in orbit about the earth. There, guests will enjoy the ultimate novelty—low-gravity for everything from calisthenics to sex.

# The Future of Plants

In the past ten thousand years, the world of terrestrial life has been more influenced by the growing population of human beings, and by their growing ability to alter the environment to suit their short-term needs, than by all other factors combined.

Through the agency of human beings, for instance, those plants they find useful to themselves have been multiplied in numbers and in living space by a factor of millions. Since there is only so much land on Earth and since all suitable portions of it were occupied by vegetation of one sort or another at the time agriculture was developed, the spread, over millions of square miles, of grain and vegetable fields, orchards, sugar cane, and rubber trees has meant a shrinkage of the forests and, generally, of the area given over to plants in which human beings are not particularly interested.

On the whole, then, the world of plants has grown increasingly unbalanced and decreasingly diverse through the agency of *Homo sapiens*.

What may we look for in the future?

If population continues to increase, energy sources to decrease, international hatreds to grow more intense, the capacity of our leaders to make wise decisions to diminish, then this trend for the plant world will continue even more markedly over the next few decades until civilization crashes. If the crash comes without a thermonuclear war, the plant world may then slowly return to its own over the remains of a shrinking humanity.

It is not pleasant to think of this particular possibility.

However, it may be that, faced with the rapidly growing crises of the closing decades of the twentieth century, humanity will learn to cooperate and take measures to control population, conserve energy, increase the humanity, rationality, and wisdom of governments, work for a better ecological balance, and encourage the continued advance of science and technology, using its products more wisely than in the past.

All this seems rather Utopian, but it may happen. In that case, what of the plant world?

The great change for plants on Earth could then well be the addition of a number of species (now more or less disregarded) to the list of those that are coddled and cared for by humanity. These may include new tropical species capable of producing carbohydrates suitable for human composition. To an even greater extent, it may include plants that yield oily saps that can be refined into hydrocarbons or alcohols that may be used as relatively cheap, nonpolluting liquid fuels.

Even greater and more startling changes, however, may take place not on Earth but in the new habitats that humanity will establish in space. The political-economic urge into space may be powered, to begin with, by the need to obtain solar energy for the world. If solar stations are to be placed in orbit about the earth in requisite numbers, then mining stations will have to be established on the moon and smelting operations will have to take place in space itself. To supply the manpower to do the work, space stations will have to be established, and these will eventually be large enough to hold anywhere from ten thousand to ten million people in permanent occupation.

The space settlements, if we follow the vision of Gerard O'Neill, the Princeton physicist, will be engineered to be as earthlike as possible. They will rotate at speeds designed to set up a centrifugal effect that will produce the equivalent of earthly gravity on the inside surfaces of the settlements. They will be lighted by sunlight reflected from a mirror and shining through louvred slats that can be opened and closed to produce day and night. There will be a normal atmosphere, which in the larger settlements could produce clouds. There will be soil underfoot that could be made rugged and even mountainous in spots. There will be housing, small lakes, rivulets—in short a world in miniature, including plant life.

Each space settlement would be engineered to reflect the tastes of its inhabitants and each might be different, but we can suppose the vast majority, perhaps all of them, will be designed to reflect a mild environment. Furthermore, the plant-world portion of the settlements would consist largely of those species useful for food—grains, vegetables, berries, fruit, and so on.

There will surely be animal life as well, primarily those animals that can feed upon plant life other than that intended for human consumption. Naturally, the animals will be intended for food, too.

It is unlikely, however, that every living thing on such space settlements will be chosen merely for its value as potential food. Man does not live by bread alone. There is bound to be space reserved for plants that will serve an aesthetic purpose. There will be gardens, patches of wildflowers, even here and there a tangle of wilderness.

In fact, each space settlement will find that in order to produce a system of living things that will flourish and remain in long-term balance, some species will have to be introduced for no direct human use at all in either the culinary or the aesthetic sense, but merely in order to occupy some environmental niche that must be filled.

It is quite possible that the coming of space settlements will initiate a new study of mini-ecological balance; the science of determining just what combination of a limited number of species will suffice to establish a stable system of life.

It would be absolutely necessary to include those species needed for a basic food supply, of course; but beyond that there may be millions of relatively simple combinations that would do, and every space settlement would try to choose one that would meet the tastes of its particular inhabitants.

It may well be—it is even likely—that no two space settlements will have precisely the same combination of species included in their plant world. Evolutionary change may cause the life-systems to grow increasingly different in detail with time.

It will all represent an added incentive for travel and tourism, by the way. Inhabitants of one space settlement, knowing their own limited world of life very well, would find each different settlement a wonderland of strange flowers, insects, rodents, and so on.

On the other hand, there would have to be careful inspections before entry. Obviously, no one could bring in "exotic plants"; but, in addition, there would have to be certainty that there were no seeds or spores unwittingly carried in the baggage or on the clothing.

Visits from Earth would be a particularly troublesome problem. The riotous macro-ecological system on Earth would ensure the constant risk of entry of one or more unwanted species. These would be unwanted not necessarily because they were intrinsically harmful, but because they would upset the ecological balance of one of those finely honed space-settlement life-systems.

On the other hand, Earth would be a tourist's paradise for the space settlers, if only because of the vast multiplicity of life-forms they could see and admire.

An entirely different class of plant-system might be developed on another kind of space settlement, one given over entirely to unicellular forms.

We can imagine such a settlement to consist chiefly of a large vessel of water, illuminated intermittently by reflected sunlight from all angles, an illumination that would keep it at some appropriate temperature. The vessel would contain algae that would multiply more rapidly than any form of multicellular organism, plant or animal, possibly could. (The vessel may have to be double-walled with clear water between to absorb hard radiation that could be harmful to the cells.)

At one end, carbon dioxide and various mineral nutrients would be introduced; at the other side, the water would circulate and the algae would be filtered out. Proper desiccation and fractionation of the algae would produce a fine nourishing meal that could serve as animal feed at the very least and, very likely, as additives for human food as well.

The whole would be automated and would consume only solar energy, essentially inexhaustible.

With more and more of these in orbit, the pressure for farmland on the settlements may be eased, and ecological balance would then be attained with a more even distribution of species.

As far as Earth itself is concerned, space will become the major source of almost all human needs — minerals from the moon, energy from the sun, and food from the churning algal cells endlessly multiplying ("pie in the sky" at last).

Must we have intact cells to produce food, for that matter?

Chloroplasts (the chlorophyll-containing granules within plant cells) could be isolated and put to work on their own, performing their task more rapidly and with more simple-minded efficiency, perhaps, than ordinary cells could.

And if the chloroplasts are studied in detail, the chemical system may come to be so well-understood that it will become possible to mimic their workings in systems simpler still.

In short, we might be able to produce food through the agency of non-living chemicals. It is doubtful that this could be quite as efficient as the complex living systems finely honed after billions of years of evolution. The nonliving chemical systems, however, would be less vulnerable to radiation, perhaps, and more easily adjustable to various conditions, such as higher or lower temperatures, or different wavelengths of light.

Agriculture as such may eventually pass away. Food production may become a matter of chemical technology.

Will we, then, no longer need plants?

Nonsense! Again — man does not live by bread alone. The plant world will always delight us for a variety of reasons; and if the amber waves of grain become less important, there will be more room for pleasure in variety. Even small space settlements may then find they can develop and care for quite complex ecological balances and may find themselves rejoicing in a diversity of plant species impossible under earlier conditions.

# 48

# Bacterial Engineering

Genetic engineering is not really something new. Human beings have been fiddling with genes for as long as ten thousand years. That's how long they have been growing plants and herding animals.

Of course, in earlier times human beings had not even heard of genes, but it made reasonable sense to see to it that an unusually strong bull sired many calves and that an unusually good milker mothered as many as possible; and, again, that one used the best strains of wheat for seed—those that grew fastest or that yielded particularly plump grains.

In consequence, over the generations the plants and animals that humanity had domesticated came to change their characteristics in the directions human beings deemed desirable. Horses are bigger, stronger, and faster than their prehistoric progenitors; cattle are more placid and yield more beef and milk; sheep yield more wool; chickens lay more eggs; turkeys have larger breasts; and so on.

Animals can be bred for amusement, too. Think of some of the breeds of dogs and pigeons that exist.

Nor does anyone ever consider what is good for the animals themselves. Many of them could no longer exist in the wild without human care. The corn plant could not even reproduce without human help.

Although human beings were not able to control the mating of smaller and simpler creatures in the same way they could the large plants and animals they had domesticated, they did what they could to make use of their labors. They plundered beehives for honey and certain caterpillar cocoons for silk. They even put yeast cells to work fermenting fruit juices and soaked grain.

All these things were done in prehistoric times. Oddly enough, humanity in historic times has not succeeded in truly domesticating any new species of plant or animal. Perhaps that is because there has been no necessity for doing so. What has already been done seems to be enough.

Yet there are directions in which we are making progress, and can make more in the future, directions undreamed of in earlier times.

There are tiny organisms whose existence was not discovered until recent times — 1,500 different species of bacteria. A few of those species cause disease, but the percentage is low. Most bacteria do us no harm, and many of them are useful, even essential.

Biochemically, the bacteria are amazingly versatile. There are no naturally produced materials that cannot be broken down by one type of bacteria or another. Decay bacteria restore everything to the biosphere to be used over again. If their work stopped, the world would become littered with undecayed scraps of indigestible matter, and these would accumulate till all life stopped.

Other bacteria can combine the nitrogen of the air with other elements to form substances that maintain the fertility of the soil. Without them, the soil and waters of the earth would slowly grow sterile.

Bacteria can carry out chemical reactions that higher animals cannot, and we sometimes benefit by it. Bacteria in the stomach of cattle digest the cellulose of grass and hay (the cattle cannot do it all by themselves). The products of digestion are absorbed by cattle and eventually come back to us as meat, milk, cheese, and butter. Bacteria in our own intestines form some of the vitamins we can't make for ourselves.

The question is: Can we tame bacteria? Can we carefully breed them into strains that are of even greater use to us than they are in nature?

We can't do it by the ordinary breeding methods we use with larger animals, of course, since bacteria reproduce asexually. However, might we not isolate various bacteria of a particular species, allow each to reproduce, then test each batch for some useful function, pick the one that performs most satisfactorily, concentrate on that for future study, and destroy the rest? Doing this over and over might result in domesticated bacteria (so to speak) that are an improvement over the natural strain, at least from our own selfish viewpoint, as is true of cattle, horses, sheep, and swine.

We can indeed do this, but we have learned even better techniques.

The genes of all organisms, including bacteria, are made up of molecules of "deoxyribonucleic acid," usually abbreviated DNA. The DNA in any cell guides the formation of enzymes that, in turn, dictate the kind of chemical abilities the cell has.

The DNA in any particular cell is passed on to new cells every time cell-division takes place, but sometimes random changes ("mutations") are introduced into the DNA when it is multiplied. When we develop new strains of organisms (whether camels or bacteria) by breeding or by selection, we are taking advantage of these mutations.

However, these mutations *are* random. They can't be counted on. If we wait for a mutation to bring about a particular desired change, we might wait for years, or we might wait forever.

And, as it happens, we have now reached the point where we don't have to deal with genes only through the organisms. We can manipulate the DNA directly. Scientists have learned how to break the long molecule of DNA at specific points and how to recombine them in new ways, or how to insert new pieces, or clip out old ones. Portions of a DNA molecule from one species can be inserted into the DNA molecules of another species. A piece of human DNA can be inserted into a bacterial DNA, for instance.

This splitting and recombining of DNA is referred to as "recombinant-DNA" research.

What are the uses of such research?

Consider diabetes, for instance, which is quite a common disease. Diabetics, at some time in life, through an unfortunate inheritance of a defective DNA molecule, lose the ability to form insulin, a hormone that is essential to the proper handling of starches and sugar in the body. Diabetics can, however, live out normal lives if they are supplied with insulin for periodic injection into their bloodstream.

Insulin for this purpose comes from the pancreases of domestic animals—one pancreas per slaughtered animal; so the supply is limited and cannot easily be increased.

But what if we design a bacterium that can manufacture insulin? We could then grow unlimited quantities of it. In fact, by inserting into bacterial DNA the proper bit of DNA from human cells, we can have the bacteria grow human insulin, which is slightly different from the insulin of cattle and swine. The latter do the work, but the former would do it better.

This has been done, at least in the laboratory. Bacteria have been produced that are capable of manufacturing human insulin.

We might, in similar fashion, design bacteria to manufacture other hormones; or to produce certain blood factors needed to clot blood, factors hemophiliacs lack; or to produce vaccines for use against still smaller organisms, the viruses.

We might also design bacteria that combine the nitrogen of the air more efficiently and rapidly. We might even improve their already versatile scavenging activities.

Suppose bacteria are developed that are exceedingly efficient at absorbing and metabolizing hydrocarbon molecules. They could be used to mop up oil spills—not only removing them from the environment but converting them into protein, which, after a number of stages of eating and being eaten, will reach our own table.

Bacteria might also be developed that can break down plastics that have been properly treated before being discarded. Plastics and other synthetics might not then act the part (as they do now) of undigestible, undecayable remnants cluttering the world.

Bacteria might be able to collect and concentrate traces of metals from wastes or from sea water and become the miners of the future.

Most important, perhaps, the work on recombinant-DNA, on forming new genes and observing how they work, may give our scientists new insights into what goes on inside the cell—insights they might not gain as easily, as directly, or as quickly in any other way.

Such insights may enable them to learn, for instance, what goes wrong when a cell becomes cancerous, and perhaps then to keep it from going wrong or to set it right after it has gone wrong. It might increase our understanding of various degenerative diseases, such as atherosclerosis, arthritis, prostatitis, and kidney failure, opening the way to prevention and treatment.

In fact, it may help us tackle senility and old age as diseases to be prevented, rather than as inevitables to be endured.

These are the advantages, the upbeat outlook on the subject of recombinant-DNA research. Are there disadvantages? If so, what are they?

In breeding cattle, sheep, chickens, and so on, we are not likely to develop strains that are dangerous to human beings. And if we do, we could surely handle it; for the first domestic animal that showed signs of ferocity in a way we found uncomfortable, would be marked for slaughter.

The situation is different in the case of bacteria, some of which live in our gut or on our skin and which can, at times, and in certain cases, produce disease.

On the whole, as I said, most bacteria do not live with us or on us, are not parasites, do not cause disease, and do us no harm. Even the bacteria that are in or on us are generally benign and cause us no serious trouble. We are adjusted to them and live with them more or less in harmony.

The adjustment is not, of course, a matter of deliberate choice, but is the product of the cell chemistry that gives us efficient defenses, both cellular and molecular. These defenses have evolved, hit and miss, because people with inefficient defenses have fallen prey to disease more often and have left fewer descendants than did those who happened to have more efficient defenses.

However, bacteria, and other disease agents, too, do undergo mutations. Sometimes a strain appears, naturally, against which our defenses do not work well. As a result there can be an epidemic and millions may fall sick, or even die, before a new balance is set up.

The many new strains of influenza are an example of this, and the worst of them in recent history was the "Spanish flu" that swept the world in 1918, just as World War I was ending. In a year, it killed thirty million people, more than all four years of World War I combined. The Spanish flu killed one-sixtieth of the world population at the time.

In the fourteenth century, a new strain of plague devastated the world. It was the so-called "Black Death" and it managed to kill one-third of the world population, according to some estimates.

Well, in recombinant-DNA research, we are forming new strains of bacteria. What if, inadvertently, without even knowing it, a strain is formed that is much more virulent than the parent strain? What if a bacteria that does not cause disease at all is converted into a killer parasite?

Might we unwittingly unleash a super–Black Death that would kill off all humanity, or completely destroy some other organisms that are important to us or that are a vital portion of the ecological fabric?

We needn't even think of killer diseases. One of the common bacteria used in recombinant-DNA research is *E. coli,* an ordinary inhabitant of the human gut, a disregarded tenant of the human body that causes us virtually no trouble.

What if a strain of *E. coli* were formed that was just a trifle less well adapted to us and we to it, and what if it got loose? It would certainly find its way into the human intestines and would flourish there, but might be sufficiently irritating in some way to produce nausea and diarrhea. We might imagine a virtual worldwide epidemic of "Montezuma's revenge." People might adjust to the new strain after a time, but even so it would be a horrible episode in world history.

Scientists engaged in recombinant-DNA work are aware of the dangers and take extraordinary precautions to prevent contamination of the environment with their products. They also point out that the new strains they develop are deliberately weakened to the point where they cannot survive except in the specialized environment supplied by the researchers.

There is, in fact, very little chance that a dangerous strain will be produced; and very little chance that, if one was, it would be able to survive on its own; and very little chance that, if a dangerous strain was produced and could survive on its own, it would be able to get out. Combine all those very little chances and the end result is a virtually zero chance of harm.

Virtually zero is not, of course, actually zero. One can imagine all kinds of weird coincidences that will end by loosing a superplague. Can we chance it? Can we do anything that will permit a chance, *however small,* of destroying humanity? (The same problem faces us with respect to nuclear weapons, nuclear wastes, and biological warfare.)

It is easy to say, "Play it safe," and to suggest that all recombinant-DNA work be stopped. If recombinant-DNA work were being carried on only to amuse scientists in their idle hours, then certainly let's stop it.

Recombinant-DNA research has, however, the potential of great good, as I have pointed out, and one has to balance this potential good, which would seem to be one whose fulfillment has a high degree of probability (considering the work already done on insulin formation) against the potential for dangers, which would seem to be one whose fulfillment has a very low degree of probability.

What do we do?

My own feeling is that the potential good of recombinant-DNA research is too great for us to abandon it but that the very natural fear on the part of many toward the potential harm (however unlikely) of recombinant-DNA is too real a factor to ignore.

A possible solution to this agonizing dilemma is that, when we properly develop a space capability over the course of the next century, we will have laboratories in orbit about the earth. (In high orbits, be it noted, that will be permanent, not low orbits—as in the case of Skylab—that will involve eventual re-entry and crash.)

In those space laboratories, potentially dangerous experiments can be carried out under conditions in which the benefit will diffuse to all humanity, while the dangers will be borne only by those working in the laboratories, who will of course be aware of those dangers and will, for rewards that seem of importance to themselves, have volunteered for the task.

# 49

# Flying in Time to Come

In order to predict the future of aviation, we must first ask ourselves what the future of society will be.

Suppose, for instance, that over the next generation the stresses that are now increasingly plaguing society force a breakdown of civilization. It may be that aviation will then dwindle into a forgotten art, with only some rusty planes and overgrown airports to bear evidence that once humanity could fly.

But that is not an interesting prediction (even though, just possibly, a true one). Let us instead assume that civilization will survive, and let us draw a possible blueprint for survival.

First, population growth will have to cease or there is no way of surviving as a civilization. Eventually, population density will have to be, if anything, somewhat less than it is now.

Second, if civilization survives, technology will continue to advance, and one obvious direction of such advance is in the improvement of communications. Imagine more and better communications satellites in space; satellites interconnected by laser beams that have millions of times the capacity for voice and picture channels that present-day radio beams have. On earth, laser communications will be carried through optical fibers.

With a combination of satellites and lasers, an enormous revolution will take place in transportation, too. It will no longer be necessary to transport mass in order to transport the insubstantial information the mass contains. This information can be sent instead, and at the speed of light.

With room in the laser beams to give every individual on earth a separate television channel, closed-circuit television will come into its own. Images of people can be sent in place of the people themselves and conferences can be held holographically in which five people, each on a different continent in reality, can sit in one room—each one seeing four images in complete fidelity.

Offices can be run, factories supervised, machinery controlled, all from a distance, through combinations of automation and television, as easily as on the spot.

The emphasis will therefore be on decentralization. No longer will it be necessary for humanity to huddle into immense swarms because that is where jobs are and where culture is. Jobs and culture will be in reach wherever we are.

What will aviation be like on such a low-density, decentralized world? Since there won't be vast population centers to feed travelers into huge airliners, and since the need for business travel will decline, the days of mass air-transportation will pass.

On the other hand, individuals living in greater physical isolation than today (but in close communication and cultural contact with others) could maintain efficient *physical* contact with the world only by flying. To visit friends in the flesh, to go sightseeing—to travel for pleasure in other words—by any other means but flight would, in a decentralized world, be too limiting.

But to serve small groups of people, we would require something analogous to the automobile. Flying flivvers may be the aviation wave of the future.

The simplest form of personal flight would be nothing more than a reaction motor strapped to a device that is chair-shaped. On the ground, motorized transport is simplest in the form of a motorcycle, two wheels to straddle, a seat, and an engine. In air flight, we don't need the wheels—just a seat and an engine.

The flying chair, if that's what we want to call it, would not, of course, have to be open to the air-stream. A light plastic front, hinged for opening and closing, will protect the traveler from the wind without obscuring his vision. The upper part of this wind-break could be tinted, or polarized, or both, to cut down sun-glare.

There would be no problem in heating the flying chair since some of the heat of the rocket exhaust could easily be diverted for that purpose. Vents in the wind-breaks, which would in any case be useful for fresh air and to prevent interior fogging, could be widened for cooling in hot weather, though outright air-conditioning could be added, too.

There's the problem of safety, of course. Any of a number of sudden defects would mean an inevitable fall. This need not be a very fearful possibility, however. The flying chair could be bottom-heavy, and bottom-padded. In addition, a parachute could be released, or fabric could be expanded into a balloon by a small container of compressed hydrogen or helium, either safety device coming into play automatically at any sudden decrease in height.

It is very likely that the safety record of the flying chairs would be much better than that of our present-day automobiles.

Where ordinary flying-chairs, built for sedate travel, would be safe and conservative, not so, necessarily, for those versions that are stripped down and designed for power and maneuverability. Those would be used by those who prefer excitement. They would be the "hot-seats."

The flying chairs, whether in one-seat or two-seat versions, would be primarily useful for short flights, the equivalent of trips to the grocery store or visits to the neighbors. Neither their speed nor their range would be remarkably high, nor, of course, would be their storage capacity.

For long trips, for vacations, for family outings to the Taj Mahal, complete with baggage, something more than a flying chair would be needed, something more like what we would recognize as an airplane.

However, assuming that a sizable percentage of families had planes, it would be wasteful to build extended runways by the millions. The trend would be toward vertical takeoffs and landings (VTOLs).

Undoubtedly both takeoff and landing spots would have to be specially designed to withstand the shock of departure and arrival, and not any place would do, but in the end such VTOL-ports will dot the world and there will be arrays of them at tourist spots.

Long journeys through the air for the average family are, in some ways, more complicated than similar journeys on land. There would be no recognized roads or road-signs; nor can one expect every flyer to achieve the art of the master navigator.

Still, we need not expect to see the VTOL-flyer leaning out the window to try to recognize landmarks below (or oceanmarks).

It would seem almost inevitable that the VTOL-flivver of the future will be thoroughly computerized. The destination, in terms of some key pattern equivalent to latitude and longitude, can be punched into the VTOL-computer, as well as that of the starting point. We can imagine the vehicle brought up manually and then turned over to the computer, which could guide itself by the aid of navigational satellites in space.

It would not be until the destination was in sight that the flyer would feel the need to move the VTOL-flivver into manual again, and maneuver to a landing.

In the decentralized world I am describing, one in which electronic communications will be ultimately sophisticated, it is not at all likely that there will be huge concentrations of aircraft along the airlanes. Yet, even so, there may be the occasional chance of mid-air collisions due to human error if not with other planes then with natural obstacles.

To reduce the chance of this to the level of computer-error (which, we can hope, would be considerably lower) may not be difficult.

Each VTOL-flivver and, for that matter, each flying chair, can be equipped with a radar device designed to detect solid objects in its path, from large birds to fog-hidden mountains. The computer will be designed to take evasive action when the radar indicates an obstacle, returning to course only when the way ahead is clear.

(One can imagine a plane feeling its way around a mountaintop in the fog, bit by bit. This could be hard on the fuel supply and might make for an unlooked-for landing in some desolate area. In the communications-saturated society of the times, however, such isolation need hold no terrors. Each VTOL might have its characteristic wavelength, sending out a signal that could be picked up, if necessary, anywhere in the world by relay through communications satellites, and which would, at once, indicate the position and even the identity of the stranded party.)

We can suppose that it would be wise to have the computer take over, on advice of radar, even when the plane is in manual. One can imagine this as the source of a new kind of game of "chicken."

Those who enjoy the excitement and thrills of hot-seat drag-races might deliberately arrange to take multiple collision courses. As the flying chairs race one another, they will be radarized out of human control, veering and curving sharply with a computer-skill a human driver could not possibly duplicate. Presumably all the hot-seats would be gotten out of the melee safely.

This "guaranteed" safety would not spoil the fun. In the first place, a driver could be shaken up badly and even hurt as the flying chair jerks through the melee. It would take a certain skill, therefore, to choose a course into the melee that might minimize the jolting, and a certain stamina to survive what jolting did occur. Both of these qualities would be highly regarded by the young hot-seaters (whom, somehow, I picture as wearing imitation-leather jackets and chewing plasticine).

Of course there is always the small chance that a computer may fail, that controls may jam, that a particular combination of courses may lead to unevadability, and there may be even fatal accidents. The chance of that will very likely increase the thrill.

These personal jets, both short-range and long-range, are high-energy means of transportation, of course, and we are now living at a time when energy, particularly oil, is rapidly running out. Even allowing for lower fuel needs in a decentralized, image-communicating world, where will the fuel be coming from to power these planes of the future?

Since we are picturing a future in which civilization has been preserved and in which technology continues to advance, we must assume that new sources of energy will be found. The two large sources that will come into use (if humanity proves intelligent, resolute—and lucky) will be nuclear fusion and direct solar energy.

Either or both could supply electricity primarily, but this could be turned into other forms of energy as well. With the use of energy, carbon dioxide and water could be turned into a hydrocarbon mixture that could be used as fuel — with oxygen left over as a side-product. (This is very much what the green plants accomplish with the energy of sunlight.)

There is no fear of using up carbon dioxide and water in this fashion, not only because the earth has so large a supply of both, but because when the fuel is burned it combines with oxygen to form carbon dioxide and water again. The process moves in a circle and nothing is used up but fusing nuclei and sunlight, both of which can last for billions of years.

Another advantage is that if carbon dioxide and water are the only raw materials, the fuel produced contains no nitrogen, sulfur, or other atoms that produce air-pollution. The air vehicles of the future will be clean.

The flying chairs and VTOL-flivvers are, of course, air-breathing vehicles that carry only fuel, and which use the surrounding atmosphere as the oxygen source.

Suppose, instead, that liquid oxygen is also stored aboard in an air vessel. In that case, the vessel could travel up into the stratosphere and beyond. Naturally, such a vessel would have to be considerably more elaborate than a VTOL-flivver. It would have to be stronger, be air-tight, carry more elaborate life-support systems, and so on.

Nothing of the sort need frighten us, though, since we have rockets right now that are elaborate enough to take men successfully to the moon and back. In the future we are describing, rockets will be much more compact and, thanks to technological advance and mass-production procedures, will be much cheaper.

They may not be so cheap that the average family will be in a position to own one. They might, however, be hired for community use. For long-distance travel, great speeds can be obtained in the airlessness of near space, and any point on the earth's surface would then become no more than an hour removed from any other point.

Then, too, near space itself can become a prime tourist attraction. Suppose a rocket ship carries a group of vacationers up into orbit (in a gradual rise that avoids too great an acceleration at the cost of expending more of the earth's essentially endless supply of fuel). There would, in that case, be the panoramic view of Earth, something of a majesty impossible to duplicate in photographs or even in television views. There is, secondly, the sensation of zero gravity, where even approximations cannot convey the feeling.

It is quite likely, too, that, if mankind is to pin its faith on solar energy, power stations will be built in space rather than on Earth (with raw materials brought, perhaps, from the moon). In space, much more solar energy is available, since there is neither night nor atmospheric phenomena to interfere with the collection.

To build such power stations will surely involve the development of an entire space industry, and space settlements to run that industry.

We can imagine the moon's orbit filled with dozens of space settlements (cylindrical, spherical, and toroidal in shape), and in that case new varieties of transportation become possible.

Between one settlement and another, spaceships could freely travel. They will be doing so on a nearly equipotential line; that is, they will not be moving away from Earth to any extent. (They may be moving away from the moon but that would represent only one-eightieth the problem.)

This means that such traveling will consume little energy. It will be like gliding on ice; you just give yourself an initial shove and move right along. Under such conditions there may be "rocketing chairs," somewhat more elaborate than the flying chairs on Earth, but still essentially one-man vehicles.

Within the settlements, however, comes true flying at last.

The settlements will turn so as to produce a centrifugal effect that will mimic normal gravity on their inner surfaces. Nevertheless, whatever the shape of the satellites, the closer one approaches the axis of rotation, the weaker the sensation of "gravity."

Eventually, close enough to the axis, the sensation of weight is so small that a person can use his own muscle power to maintain himself in the air, if he makes use of artificial wings that are at once light and strong.

We can imagine human bats, so to speak, maneuvering, dipping, soaring, and racing in the air of the space settlements. And that is the ultimate, for that is the true dream of humanity — to fly by unaided muscle power and to be freed from the prison of surfaces without being enclosed in other surfaces.

And all of this might happen.

---

## 50

# The Ultimate in Communication

If it is the ultimate in communication we are seeking, we already have it, in one sense at least. Indeed, we attained it over a century and a quarter ago.

In 1844, the first telegraph line was put up between Baltimore and Washington, and across the wires winged the first message, "What hath God wrought?" At that moment, information was transferred at the speed of light, 186,282 miles per second, over a sizable distance.

This speed of information transfer has not been exceeded since and, physicists are quite certain, can never be exceeded. What we have done since has been to add refinements.

The telegraph sent messages in code, but the telephone, invented in 1876, transmitted actual words. Both telegraph and telephone extended themselves in space over uncounted thousands of miles of wire and cable. Radio, however, in the 1890s, transmitted information on electromagnetic waves alone, doing away with the necessity of wires (hence, the "wireless" as it is much more appropriately called in Great Britain).

Radio transmitted only code at first, but by 1906 it was transmitting words and music. Television, going commercial in 1947, added visual information, so that we got images as well as sound; and ten years later, color was added to the images.

Now television rules the world, transmitting sound and full-color image at the speed of light.

What can we possibly add to that?

To answer that question, we have to understand some of the shortcomings of ordinary television. The electromagnetic wave bands used for ordinary commercial television have room for relatively few channels. Then, too, the

278

beam carries only to the horizon, for it moves in a straight line and cannot follow the curve of the earth.

Because of this, hundreds of local television stations are required to cover an area the size of the United States, and any given spot has available to it only a limited number of channels.

Because television transmitting stations represent a huge capital investment and require expensive maintenance, they cannot make money unless they receive large sums for the product. These sums come from advertisers who must reach huge audiences to recoup *their* expenditures. Since huge audiences are available only for the sort of thing that will not cater to special tastes, only run-of-the-mill television fare (with rare exceptions) is possible, and no amount of breast-beating will change that fact.

The next advance, then, must be to short-circuit the transmitting station and supplement commercial television with material that will cater to minority viewers. (It is important to remember that we are all minority viewers. No one is *entirely* a common denominator. The most average man in America will have some likings and interests not shared by the majority.)

Cable television improves the quality of the picture from commercial stations and, in addition, offers what is, in essence, special fare, for pay. Payment by the consumer can make advertisers unnecessary (where there are no large expenses involved through high-priced entertainment personalities or difficult special effects). If there are no advertisers demanding viewers in the tens of millions, it may then be sufficient to satisfy smaller numbers of viewers who, after all, have their rights too.

Indeed, if we pass on to TV-cassettes, we sell not a program in progress, which goes winging its way out on microwave beams to be seen at that moment or perhaps not at all, but a frozen program on tape. It is the tape that is sold—in numbers that may be in the millions, or the thousands, or the dozens—and it is the tape that is played on a television tube privately and at the convenience of the viewer—once, twice, or a hundred times.

When this is done, television becomes completely democratized. You get what *you* want; you don't merely participate in what nearly everyone is willing to endure.

If someone wants to watch an illustrated lecture on chess with animated sequences, or a handball contest, or fancy dives, or some news on cancer research, or demonstrations of yoga positions, or a stag movie, for that matter, he will not and cannot count on having it on commercial television, but he can have it on a cassette.

More and more, television in the future will grow to resemble, in image and word, what the publishing industry is in word alone. Cassettes will be "published" in all categories, from best-sellers to prestige items, from juvenilia to special fiction, to nonfiction, to textbooks.

The printed word will not be replaced or grow less important (but then the printed word has always been the refuge of a small minority—few Americans read as many as one complete book a year), but it will be supplemented.

For instance, a newspaper is still superior as a transmitter of news, by a nearly infinite margin, to the average television news broadcast (which is usually a reading of those headlines that lend themselves to image-illustration), but who says that a newspaper must be printed on a forest of woodpulp and delivered in pound-lots to individuals?

It can be transmitted by screen in a fashion so controlled that it can be skimmed, or halted for closer reading, with particularly interesting items—the financial page, for instance, the sports page, the comic page, a certain news story—printed off on demand.

Again, there is a democratization. You get exactly what *you* want, not everything that everybody wants.

The technology of communication can advance with respect to distance, too. As I said before, the speed of electronic communication is such that, in theory, all parts of the earth can be reached from all other parts in negligible time.

The difficulty, in practice, lies in following the curve of the earth. This has been done by using wires and sending signals along them; but hundreds of thousands of miles of wires in all directions are expensive to produce, and lay down, and maintain. Radio waves do away with the wires but require tricky reflection from the upper atmosphere and there is limited room in the carrier waves for different messages.

The result is that, despite the speed of electronic communication, it costs more to send messages across long distances than across short ones.

The answer to this problem is the communications satellite, circling the earth at a height of 22,300 miles above the surface. At that height, a satellite must complete a revolution about the planet in just twenty-four hours and therefore (to an observer on the earth's surface) would seem to be hovering over one spot on the equator. (If the satellite is not positioned over the equator, it will seem to move north and south in various complicated patterns.)

Three or more such satellites, properly placed, will blanket the earth. One more will be above the horizon as seen from any spot on the earth's surface.

A message sent to a satellite and relayed back (perhaps via a second satellite) is distance-insensitive. The satellites have all of the earth in their view and it is no more difficult to reflect a message to a spot on the other side of the earth than it is to send it to the place from which it is being sent. It therefore becomes no more expensive or difficult for a New Yorker to dial a number in Samoa than to dial one in New York.

To connect the satellites and the earth, laser beams may be used. These are made up of light waves, which are far shorter than microwaves; so there are a far greater number of light waves of different lengths than there are of microwaves. Modulated laser beams would leave room for many millions of voice and picture channels. (Laser beams are ideal for transmission through a vacuum but not so good for transmission through air — some clever engineering will be required.)

Communication satellites will produce a kind of withering of the earth-bound hardware required for communications. There will be enough wavebands available to allot each individual (we might speculate) a personal wavelength. By tuning to a specific other wavelength, he can be in communication with any person on the face of the earth. As far as communications are concerned, the whole planet can become a village.

In fact, no one can possibly get lost while his personal telephone is working. An "I am lost" steady signal can lead to his location, wherever he is.

With communication totally democratized in this fashion, the technological basis for world government will be set. With no sensation of distance and with everyone here and now with everyone else, national boundaries will make less sense than they do now (if possible). The world will seem tiny, with the problems faced by one nation easily made clear to another (or freely discussed, at any rate). The push toward the adoption of a kind of worldwide lingus franca, probably based largely on English, will be great.

With distance defeated, communication will continue to replace transportation. Until recently (as history goes) the only way one could transmit information was by transmitting mass. Messages went by way of runners, couriers, and letter-carriers.

Yet what is important is the message and not the mass. If information can be transmitted by massless electronics, is that not so much the better? We have witnessed movements of this sort in the past. The telephone has brought about a great decay in the art of letter-writing, and why not?

Add to the transmission of words the electronic reproduction of facsimiles of reports and documents and there will be a further reduction in the actual transferral of mass.

Even human beings can move from place to place by image rather than by mass. With communications satellites and laser beams, there will be enough channels available to allow closed-circuit television in almost any reasonable number and spanning any distance at all.

Conferences can be held in which individual men, seated in five different continents, can have their images all together. If they need documents, these can be reproduced and brought from one point to any other at the speed of light; information can be supplied from a central computer to any point.

The whole world could decentralize. No one would need to be at any one particular spot to control affairs, and businessmen would not need to congregate in offices. Nor, with the advance of automation, need workingmen congregate in factories.

Men locate themselves where they please and everything — information, entertainment, trivia — can be brought to them.

This does not mean that people won't travel. They might choose to for any number of reasons. But subtract from the total movement of population those who travel only for business reasons or out of necessity and those who are left can travel in greater comfort and place less strain on the planetary technology.

So far all I have talked about is well within the limits of modern technology. Much of it is already here. Cable television, cassettes, communications satellites, lasers—all here. The development of a new world-system of communications, and of a new life-style based on such a system, is more a matter of economics, politics, sociology, and psychology than one of technology.

Can we, however, move farther into the speculative?

Suppose we visualize every man with his own projector and his own receiver—a universal videophone capable of transmitting and accepting sound and image. We would want it small and mobile, but very complex and versatile, and yet stable and reasonably foolproof. We would want it, in fact, to have the compactness, complexity, and stability of the human brain (the most compactly complex sample of matter *by far* that we know of).

But why search for something like the human brain, when perhaps we can *use* the human brain? Why not extend the usefulness of the brain (which can interpret light waves, sound waves, chemical molecules in the air and in solution, physical contact, and so on) by devising some sort of transducer capable of converting electronic information into the sort of impression the brain can receive and interpret.

How this can be done I have not the faintest idea, but I visualize such a device (comparatively simple, easily repaired, and inexpensive to replace) implanted in some convenient part of the body and hooked in to the nervous system.

By an effort of will, then, it might be possible to tune the mind to one wavelength or another and to receive sound or image directly. By another effort of will, it might be possible to transmit a message on some characteristic wavelength assigned to the individual.

(The phrase "by an effort of will" may sound mystical, but consider that at this very moment you can, if you wish, by a mere effort of will, raise your arm. What it is you do to your nerve and muscle cells to bring this about you cannot possibly describe, yet your arm goes up merely because you want it to.)

A world in which such direct communication, mediated by the brain, is common would obviously be quite different from the modern one. Assuming the existence of communication satellites, it would be very much as though everyone in the world were immediately next to you.

The question of privacy arises, of course, but that is nothing new. The telephone is an intrusive device that impinges on privacy, but you don't have to answer it; you can even, under appropriate conditions, turn it off.

The advent of a picture-phone is a further invasion. Even if you are willing to talk to someone, you may not feel in a condition to be seen—so the image-attachment need not be hooked in.

In the same way, the new communication by transducer and brain has the potential for an even more intimate assault on privacy, and this, too, can be controlled.

If by brain control you can emit a characteristic wavelength that can be picked up anywhere in the world, you can always be located. Your actual position will be known to anyone at every moment. Yet it may well be that there will be times when you will not wish to be located. It would be as possible for you *not* to send out the call as to send it.

Nor need the endless bombardment of electronic beams on the brain necessarily be maddening. It depends on what one is used to.

We are constantly bombarded with light and sound at present, and we grow hardened to it and learn to ignore it and concentrate only on what we want to receive. The endless sights and noises of a city street form a vague background, but we are instantly alerted by the passing of a pretty girl or a fancy car or the sound of our own name.

In the same way, it may be that the brain-receivers of the future will deliberately keep your mind open to the soft reception of the full range of wavelengths. You will be dimly aware of countless impingements, but they will not disturb you. But let your name sound, or something of interest, and your brain will zero in, automatically select the desired wavelength, and blanket out the rest.

With practice, it might not even be necessary to close one's eyes and eliminate ordinary vision in order to concentrate on the images impinging electrically on the brain. The two reception areas might be so different that it would be possible to drive a car, let us say, and keep one's eyes on the road while a set of images is reeling itself off in the brain—just as you can now drive a car and listen to music at the same time, or be occupied with the images of reverie at the same time.

Perhaps, though, we can go farther still.

So far, I have only imagined the transmission and receipt by the brain of electronic messages. It is the transmission and receipt of the same kind of purposeful information we now transmit and receive by speaking and hearing, by showing and looking. It is another way of producing an activity of the brain, rather than by the brain itself.

Can we do better than that? Can we, through some form of amplification and transduction, convert the actual electrical activity of the brain cells into a message that can impose itself on another brain? Can we receive and send messages that contain all the thoughts, emotions, sensations, and fragmentary associations of brains?

The word for it is "telepathy," but most people think of telepathy as the transmission of a message of the same kind that we now transmit by speech. They imagine someone thinking, "How are you?" and someone else sensing, "How are you?"—but where is that an essential advance over saying and hearing the phrase in the ordinary way.

No, if telepathy is to have meaning, it must transmit more entirely and more intimately than words can.

Since the essence of the personality, the sense of identity and individuality, rests (it seems to me) with the thought-and-consciousness patterns of the brain, true telepathy must mean the at least temporary sharing of personality. For the period of time during which telepathic communication is being sent and received, those sharing it are at once both themselves and everyone else.

Naturally, to protect privacy, telepathic communication must be voluntary and capable of being withheld, or broken off, at will.

In telepathy, we will surely go beyond communication as we think of it today. By opening the mind completely, deliberate communication of the "Have you seen Mary lately?" type would have to be drowned in the thoughts of Mary in many broken and fragmentary forms, informational and emotional mingled together, all the associations of Mary, to say nothing of flashes of incongruous irrelevancies.

It may be that to communicate in this fashion we must withdraw, mask the brain, and let only the question through.

But is ordinary communication all we want? Might it not be that the sharing of personality will represent communion rather than communication; that it will represent a kind of intimacy we cannot conceive now, or that perhaps cannot be sustained for long (any more than sexual orgasm can), but that would represent a joy for which there is no vocabulary of description now?

And it may be that in time to come people will wonder how it was in the days when individuals were permanently and forever imprisoned in their skins, forever alone, forever lost in isolation; and they will feel gooseflesh crawl up their spine at the thought and be sorry for us and merge personalities to forget.

---

## ─────────51─────────

# His Own Particular Drummer

Back in 1951, I wrote a story called "The Fun They Had." It was only a thousand words long and its plot was a simple one —

Two children of the twenty-second century find an old book that, among other things, reveals the nature of the educational system of the twentieth century. To their astonishment they discover that large groups of children once went to special buildings to be subjected to community education by human teachers.

As the younger child, Margie, returns to her own home where her own teaching machine is waiting to continue working with her on proper fractions, the story concludes:

> She was thinking about the old schools they had when her grandfather's grandfather was a little boy. All the kids from the whole neighborhood came, laughing and shouting in the schoolyard, sitting together in the schoolroom, going home together at the end of the day. They learned the same things so they could help one another on the homework, and talk about it.
>
> And the teachers were people. . . .
>
> The mechanical teacher was flashing on the screen: "When we add the fractions 1/2 and 1/4 . . ."
>
> Margie was thinking about how the kids must have loved it in the old days. She was thinking about the fun they had.

The circumstances surrounding the writing of the story were these. An old friend of mine was editing a syndicated children's newspaper page, and he asked me for a little science-fiction story. I was in the mood to try irony and I was certain that children have as keen a sense of irony as adults do.

Of course children fall short through lack of experience, so I thought I would hit them right where they *did* have experience and wrote "The Fun They Had."

Is there any youngster, I thought, who would not instantly be aware that school was *not* fun? Wouldn't he see that it was ridiculous for the child who had all the advantages of a personally oriented private education to long for the barbarism of an earlier day?

After all, I myself had gone to school once and had done very well, too. I had managed to finish high school at fifteen and I hovered near the top of the class, if not at the actual top, at all times. School was about as good for me as it could be for anyone. Yet I remember:

— The bullies who made life a misery in the halls and yards.

— The slow students along with whom you had to crawl in weary boredom (or the fast students, to look at the other side, along with whom you had to race in anxious frustration).

— The inept teachers who could make any subject dull.

— The cruel teachers who sharpened their claws of sarcasm on the backs of suffering children who were not allowed to talk back.

— The strict teachers who, dissatisfied with the innate deficiencies of a school, made it a prison as well.

— The relentless competition in marks that taught every kid he was nothing unless he could grind his fellow-kid's face into the dirt.

Do you expect children to have fun under those circumstances? Is there a child who wouldn't rather have have a television set of his own, interested only in him, infinitely patient, and adjusted to the beat of his own particular drummer?

Since "The Fun They Had" first appeared, it has been in constant demand for anthologies and has appeared more than thirty times that I know of. I would think, then, that I had accomplished my purpose effectively were it not for the fact that, among adults at least, a sense of irony seems notably absent in many cases. The comments that some of the anthologists print in connection with the story, together with certain letters I get, often make it clear that the story is interpreted nonironically as a boost for contemporary education and as an author's expression of horror at the thought of machine-education.

Apparently, there is a strong trend of thought among educators that there is something dehumanizing about machine-education.

Why, I wonder?

Is there a feeling that a machine is cold and hard and cannot possibly understand the needs of children?

Yet if every human teacher who was cold and hard and could not possibly understand the needs of children was removed from his or her position, I suspect our educational system would dissolve.

Is there a feeling that a child would not relate to a machine?

I suppose that in the early days of the automobile there were those who felt that no one could possibly relate to a cold, hard, dead machine as well as they could to a beautiful, sensitive, living horse. Yet, though there may be faults in our automobile culture, they do not arise out of any shortage of affection of human beings for the cars they drive.

Is there a feeling that a machine-educated child would not have contact with other human beings and would therefore be seriously lacking in many values?

Yet who would suggest a total substitution? There would be many fields of study that would require the mass-experience—sports, for instance, nature-study field-trips, drama, public speaking, and so on. On the other hand, there would be academic subjects that do not require companionship. In fact, one could study mathematics or history, to begin with, all the better if the artificial and unnecessary open competition of the classroom was removed. Then, at appropriate times, seminars could be arranged in which students could listen to each other, comment, and profit.

In short, machine-education is not a substitute for human interaction, but a supplement. In fact, human interaction could proceed all the better if it is not oppressed by the negative conditioning of an association with a dull and uninspired mass-educational procedure involving subject matter that has nothing to do with the interaction.

Suppose, now, that our civilization endures into the twenty-first century (a supposition that can by no means be taken for granted) and that technology continues to advance.

Suppose that communications satellites are numerous and are far more versatile and sophisticated than they are today. Suppose that, in place of rather nonroomy radio-wave carriers, it is the incredibly capacious laser beam of visible light that is used to carry messages from earth to satellite and back to earth.

Under these circumstances, there would be room for many millions of separate channels for voice and picture, and it can be easily imagined that every human being on earth might have a particular wavelength assigned to him, as now he might be assigned a particular telephone number.

We can then imagine that each child, as in "The Fun They Had," might have his own private outlet to which could be attached, at certain desirable periods of time, his personal teaching machine.

It would be a far more versatile and interactive teaching machine than anything we could put together now, for computer technology will also have advanced in the interval.

We can reasonably hope that the teaching machine will be flexible, versatile, and capable of modifying its own program (that is, "learning") as a result of the input of the student. In other words, the student can ask questions that the machine can answer and make statements and answer tests that the machine can evaluate. As a result of what the machine gets back, it can adjust the speed and intensity of its course of instruction and can shift it in whatever direction student-interest displays itself.

Nor need we suppose that the teaching machine is self-contained and is as finite as an object the size of a television set might be expected to be. We can imagine that the machine will have at its disposal any book, periodical, or document in the vast central, thoroughly encoded, planetary library. And if the machine has it, the student has it, either placed directly on a viewing screen or reproduced in print-on-paper for more leisurely study.

Naturally, education cannot exist apart from the world. Its nature must be affected by the state of society.

If civilization is to survive into the age of the communications satellite, laser-beam transmission of information, computerized central libraries, and teaching machines, we can be reasonably sure that the world will by then be far more closely knit than it is now.

Living as we do now in an age of life-and-death problems, we can see that without adequate solutions to such problems as population, pollution, scarcity, alienation, and violence, civilization will not survive. Every one of these problems affects the entire globe and cannot be solved by any one nation within its own territory.

If, for instance, the United States were to stabilize its population at a reasonable level, while the rest of the world continues to expand in numbers, then the chaos, anarchy, and starvation beyond our borders would upset the smooth functioning of world trade on which our own exalted standard of living depends. The economic strains upon us would topple us into chaos as well.

Similarly, pollution of the air and ocean at any one point on earth contributes to that pollution everywhere. We can in no way isolate our air and water from that of the rest of the world. We cannot isolate our section of the ozone layer. We cannot be sure epidemics elsewhere will not spread to us, or radioactive contamination either.

Our problems are global in nature, and they can only be dealt with globally.

Consequently, in a process that has already begun, international organizations will multiply and grow more powerful as they face those various problems. However much the world may continue to give lip-service to an outmoded, unworkable, deadly nationalism, the twenty-first century may see us effectively, if unacknowledgedly, under a world government, and local differences in culture, society, or economy won't alter that fact.

Education will have to adjust to such a world. History, for instance, will have to be *human* history, with the accent on social, cultural, and economic trends, while much of the hero–villain business of war and politics will be de-emphasized. It would, after all, be ridiculous to perpetuate hatred in a world in which there is no alternative to cooperation.

Then, too, while the local languages will be used in teaching students, it would be only reasonable to suppose that some lingua franca will have developed for a world grown small and interconnected. There need be no reason to try to suppress language—for cultural variety is a great good and contributes to the sparkle and excitement of the species—but to be multi-lingual is not so unusual. Why, then, should not everyone speak at least two languages, his or her own, and "Planetary"?

Would it not be safer to have machine education in so global a society? Could we expect human beings to be entirely free of the prejudices of an earlier and nonglobal day? And, even if machines were programmed by human beings not devoid of prejudice, the mere fact that machines could modify their programs in accordance with the needs and abilities of the students might mean that they would drift toward globality.

For instance, it seems to me very likely (though my own chauvinistic wish may be father to the thought) that Planetary would be closer to English than to any other language, since English is already the first or second language of more people on earth than any other, with the possible exception of the multi-dialect and geographically isolated Chinese language.

Even if this is so, however, Planetary will undoubtedly absorb enough of other-language vocabulary and grammar to become a foreign language to those who speak English as a native tongue. Teaching machines may be so programmed, and so self-modified, as to encourage the drift of Planetary away from English in order that it be as much as possible a new language without too obvious antecedents. On the other hand, they may be programmed to resist the too extreme breakup of Planetary into mutually incomprehensible dialects (as once happened to Latin in western Europe).

Education may undergo another important revolution in the twenty-first century—it may no longer be child-oriented.

It is natural to suppose that it is only children who must be educated. They are born with no more than a handful of biological instincts and must learn everything that makes them culturally and socially human. Once they have learned a certain minimum of what is required—to speak, to read, to earn a living—their education is considered complete. Of course certain aspects of education continue, for an adult is bound to learn new matters in fields of interest, or to become more proficient in the social graces, or to adapt to new cities, new conditions, new situations. Such adult continuation, however, is left to itself—a matter for each individual to take care of. *Institutionalized* education, is left to the young.

As a result, we think of education as closed-ended. One *has* an education; one *is* educated; one is *done* with education, and enters the "real world."

This view damages both young and old. Children quickly learn that grown-ups don't have to go to school. If there is any inconvenience to school, it is attributed by children to their crime of being young. They come to realize that one great reward of growing up is to become free of the school-prison. Their goal becomes not that of being educated, but that of getting out.

Similarly, adults are sure to associate education with childhood, something they have fortunately survived and escaped from. The freedom of adulthood would be sullied if they were to go back to the life-habits associated with the education of children. As a result, many adults, whether consciously or unconsciously, find it beneath their adult dignity to do anything as childish as read a book, think a thought, or get an idea. Adults are rarely embarrassed at having forgotten what little algebra or geography they once learned, just as they are rarely embarrassed at no longer wearing diapers.

People, as they grow old under such circumstances, generally cannot change views or attitudes to conform with changing conditions and environments. They have forgotten how to learn and must rely entirely on what scraps of dimly remembered catch-phrases they picked up as teenagers. There is, in consequence, a hardening conservatism, a growing intolerance of whatever is new, and a certain mindless refusal to adjust even where that would be to their benefit. (There are, of course, individual exceptions to this.)

The relatively young who have just escaped from school, or who are in the process of escaping, tend to despise the previous generation for their conservatism and obscurantism, and even a very few years is enough to produce a noticeable difference. That is why we had the catch-phrase of the 1960s: "Don't trust anyone over thirty."

The distrust of the young for the old and their acceptance of the stereotype of age as inseparable from dullness, backwardness, stodginess, and noncreativity helps produce and confirm the truth of that very stereotype. To the degree that the young accept that stereotype, they shrink into it as they age, and it becomes a vicious cycle of self-fulfillment.

This stereotype of age as an unprofitable excrescence on the body social is not exactly new; but, as it continues, it is becoming more dangerous and, by the twenty-first century, it may be deadly.

The reason for this is that the age profile of our society is changing. Through most of mankind's stay on the planet, a high birth-rate and a high death-rate have sufficed to keep the average age of human beings low. (The average age would be lower still were it not that infant mortality contributed a disproportionate share to the death rate.)

Since the mid-nineteenth century, however, the death rate has been dropping, thanks to the advance of modern medicine. The drop in infant mortality and the slower sag in the birth rate has not been able to make up for the effect on the age profile of an extended life-span. In nations where the effects have been most marked, the average age of the population has been increasing relentlessly.

In the United States, the steadily increasing percentage of the population made up of those over 65 has now made the oldsters a formidable voting power. What's more, we are becoming, increasingly, a nation with its finances organized about pensions, medicaid, and Social Security benefits, which are enjoyed by so many and looked forward to by so many more. As some have pointed out, there are ever more and more unproductive oldsters being supported by the labors of a smaller and smaller reservoir of productive youngsters.

And what if this tendency continues? Consider—

The total population of the world stands now at 4 billion and it is increasing at the rate of 2 percent a year. By 2010, if this growth continues unchecked, the world population will be 8 billion; by 2045 it will be 16 billion, and so on.

No one really expects the population rise to continue unchecked for very many decades, however. The only question is what it is that will check it.

It could be a rise in the death rate through starvation, disease, social strife, and so on. Checking the population rise in this fashion would, of course, produce enough misery and anarchy to shake the lofty and formidable, but rickety, underpinnings of our complex industrial society. It would shatter the technological structure that alone keeps the earth's population fed, clothed, and secure (however inadequately). With that structure destroyed, there will be no twenty-first century worth discussing.

The alternative is to lower the birth rate the world over. There are formidable obstacles to this, but as catastrophe comes closer and as a lowered birth-rate is more and more clearly seen as the only route to survival, a panicky mankind will take more and more drastic measures to ensure it; and then, perhaps, we will win through with only a minor catastrophe—that is, one from which civilization can recover.

In that case, though, a drastically lowered birth-rate the world over will ensure the continued increase of the average age of mankind. There will be a steady increase in the proportion of oldsters who will have to be supported by a steadily decreasing number of youngsters. This change will be further exacerbated by the fact that a continuing civilization will ensure further advances in medicine, so that there will be an increasingly successful treatment of the degenerative diseases that now strike the aging with such dreadful consistency—and a further drop in the death rate that the falling birth rate will have to match.

To be sure, the aged will be healthier and stronger as medicine learns to inhibit and/or ameliorate arthritis, cancer, circulatory disorders, kidney disease, and so on. To that extent, the oldsters will be less of a physical drain on society than they would be under present conditions. On the other hand, if the stereotype of the aging as mentally rigid and creatively incompetent continues to be converted into actuality, all of society will calcify. Mankind will avoid the death-by-bang of the population explosion to suffer the death-by-whimper of massive old age.

*Unless* education does something to destroy the stereotype.

Education must not any longer be confined to the young. The young must not look forward to its completion; the old must not look back on it as an accompaniment of immaturity. For all people, education must be made to seem a requirement of human life as long as that endures.

Why not? That for which a living and healthy organism is adapted is easy to do and there is no reluctance attached to it. It is no pain for a cell to divide, for a tree to put forth leaves, for a horse to run, a seal to swim, a hawk to fly. Where an animal is sufficiently advanced to allow an apparent display of emotion, it is almost inevitable to interpret its behavior as indicating outright pleasure in the utilization of its body for the purposes to which it is adapted.

For what, then, is the human body adapted? Consider the colossal human brain, making up 2 percent of the human body and weighing three pounds altogether. No other organism with the exception of the dolphin has a brain that is at once so large and yet combined with so comparatively small a body. For what is this brain adapted but for all the processes we call thought, reason, insight, intuition, creativity?

Would it not seem natural to suppose that there must be pleasure for the human being in the very act of thinking, since the brain is so adapted for it, and since its underutilization gives rise to the very painful condition we call boredom?

When a child learns to talk, it talks constantly, it asks questions, it pries and probes and is endlessly curious. It clearly loves the thinking ability it develops — and then it goes to school and has it shot out from under him.

School *isn't* fun, but might not education by machine — personalized, adaptable, versatile — prove to be fun? If it were, might it not be the kind of fun that people would hate to give up? If so, then education could continue into advanced age. Oldsters aren't likely to give up golf, or tennis (or sex, for that matter) just because they were better at it when they were younger. Why, then, should they give up education if *that* proves to be a continuing pleasure?

In fact, given a long, vigorous, and a healthy life, and one in which it is no disgrace for a mature person to "go to school," why should there not be regular switches in fields of endeavor? At age 60, why might not someone suddenly decide to study Russian, or take up mathematics or physics, or venture into chess or archaeology or bricklaying? What could better serve to keep a person's mind active and happy and alive and creative than to send it surging in new directions?

Computers, programmed with ever greater versatility and themselves increasingly capable of learning by interaction with human beings, could be ready to help further those new interests and in this way, too, teaching machines could help save our society.

The key to this vision of education is, of course, that people enjoy learning *if* they learn what they want to learn. This is not a very profound observation, actually. A child who finds every school-subject boring and incomprehensible, and who seems incapable of learning, may yet bend his every faculty to an understanding of the rules of baseball, and may succeed in memorizing, with fiendish intensity, unrelated statistics that even a mathematics professor might have trouble with.

Then why not allow a child to learn what he wants to learn? If he wants to learn baseball, let the machine teach him the academic aspects of baseball, which he can then apply in the field. He may, as a result, want of his own accord to learn to read better in order to read more about baseball, and want to learn arithmetic in order to calculate baseball averages. Eventually, he may find he likes mathematics more than baseball. Remove constraint and he may well move in a direction toward which force would never budge him.

But can the world continue if everyone has the option of choosing what he is to learn? Can society survive with an educational system that consists entirely and solely of electives?

Why not, if we take into account the likely nature of technological advance—assuming civilization survives? An increasing rate of computerization and automation of the industrial structure of society should diminish the kind of dull and mindless scut-work that now occupies so large a proportion of the efforts of humanity. One might imagine a world of robots and computers that farm and mine and tend machinery, leaving for human beings precisely the type of creative endeavor for which their brains are suited.

Every child who is not markedly brain-damaged shows the capacity to learn to walk, to speak, to adjust himself to life in a million ways even before he enters the first grade. It is clear, therefore, that the potential for creativity is present in everyone provided only that we make learning pleasant and stimulate (not penalize, as we usually do today) any demonstration of that creativity.

Let each follow his own particular drummer, and if some decide to sink into what we would think of as ignoble sloth, or to indulge in what we would now consider trivia, they may later in life grow weary and try, instead, something that our present prejudices consider more worthy—scientific research, politics and command, literature, arts, entertainment, and, of course, education. And some may move in that direction from the start (and, perhaps, abandon it later). Might it not be likely that, on a strictly voluntary basis, enough people will opt for the socially important activities to keep the world going?

And perhaps it will turn out that the activity that is of greatest importance is education — the designing of computer programs in new and esoteric directions — and of greater subtlety in the older disciplines. There could be a steady, synergistic interaction of man and machine, each learning from the other, and each advancing with the help of the other. The distinction between the two varieties of intelligence may grow dimmer, and the discovery and refinement of knowledge and the beating back of the vast cloud of ignorance may be carried on at a faster rate together than either could alone.

In such a utopian world as I describe (assuming it can be attained — which is, of course, doubtful) there is the danger that everything will run so smoothly and safely and securely as to remove all interest and produce a society that will slowly and somnolently sink into the slumber of desuetude.

Yet that need not be so, for out there in space we can find a new and more distant horizon than any we have encountered before, a greater and more dangerous frontier, a larger and more unexpected habitat, an outspreading volume with more fearful unknowns and more exciting possibilities than anything we can now imagine.

But that is a subject for another article.

# 52

# The Future of Exploration

Let us agree, to begin with, that by "exploration" we mean the venturing of a human being into a place where no human being has been before or, at the least, where no "civilized" human being has been before.

In that case, the earth is not what it once was as far as exploration is concerned. Undoubtedly, there are a few mountain peaks that have not felt the tread of the human foot, a few hidden valleys not yet visited by civilization, some obscure caves that remain as yet in hiding, and, of course, tens of millions of square kilometers of the ocean floor. On the whole, though, there is nowhere on earth we cannot travel if we but decide to take the trouble to do so.

The future of human exploration, with its full glamor and danger, rests in space; and there, too, let us agree to discount exploration by instrument. Where may we expect human beings to go?

Already an even dozen human beings—all American males—have set foot on our moon, so that exploration in the fullest sense is no longer confined to our single world. Unquestionably, we can spend a great deal of time exploring the moon in detail even if trips there and back become routine.

But take that for granted and let's ask where else we can go.

To begin with, the moon is at our back door, some 380,000 kilometers away. That isn't much even in earthly terms, for it is only eleven times the circumference of the earth, and there must be a great many tourists and businessmen who have logged far more than that in air-travel in the course of their lives. And in terms of rocket speed, the moon is only three days away. It took Columbus thirty times as long to cross the Atlantic as it took Neil Armstrong to reach the moon.

But the moon is the *only* object in our immediate neighborhood. The next nearest object that is at least as large as the moon is the planet Venus and, at its closest approach to us, Venus is 109 times as far away as the moon is. What's more, Venus is that close for only a brief period every 19 months.

At the present state of the art, a rocket ship takes at least half a year to reach Venus — and every other place in the solar system is farther off still.

Of course there are a few objects that sometimes approach more closely to the earth than Venus does. There are a handful of asteroids that pass the earth at a distance of a few million kilometers now and then, and there are occasional comets that do the same.

It might be interesting to speed toward such objects as they approach, make a landing, look about, and leave.

It would not be a good idea to hang around too long though, for all these objects are hurrying along in their orbits and eventually begin leaving the earth dangerously far behind. Nor would it be reasonable to remain on them till they came earth's way again. That is likely to take several orbital turns and therefore a number of years, and in the course of each of those orbital turns they are apt to come uncomfortably close to the sun.

For that matter, even if we decide to make the long trip to Venus, and even if our space-travel technology advances to the state where the trip can be shortened and where the spaceship environment can be made quite comfortable, Venus has a particularly poisonous atmosphere 90 times as dense as ours and a temperature that is everywhere and at all times in the neighborhood of 475° C.

We have landed objects on Venus that have made useful observations for the few minutes they endured, and we have mapped Venus's surface by radar; but it doesn't seem in the cards that human beings are ever going to penetrate Venus's atmosphere, let alone stand on its surface.

Mercury, which lies farther from us than Venus does, is not as hot as Venus is, even though it is considerably closer to the sun, because Mercury has no atmosphere to store heat. Furthermore, Mercury rotates slowly so that any given part of it experiences night for six weeks at a time and has that interval to cool off in.

We can imagine a space-traveling team landing on the night-side of Mercury and having a reasonable period for exploration before the sunrise forces them away. Nevertheless, to reach Mercury will require human beings to move in the direction of the sun, and it may be some time before space technology will make it safe for us to do that.

Surely, our first major exploratory steps beyond the moon will be in the direction away from the sun. In that direction, the nearest target is the planet Mars.

Mars has been visited by human-made objects, on and off, for sixteen years now. Space probes have skimmed by and taken photographs of Mars and its two tiny satellites. Some have gone into orbit around Mars, and two probes have landed on the Martian surface and observed it by camera and chemistry. In consequence, we know Mars in considerable detail.

There is no reason human beings, if they reach Mars, cannot remain on it for a reasonable period. Mars is not a comfortable environment for human beings but it is more like earth than the moon is. It has a tiny (if unbreathable) atmosphere, a 24-hour day, a stronger gravity than the moon, and in spots is no colder than Antarctica.

However, a round trip to Mars and back may require something like a year and a half, which is a rather rough undertaking at the present moment, and there is certainly no indication that a manned Martian trip is in the cards.

Anything farther than Mars is going to take longer still. If we talk about the outer solar system, Jupiter and beyond, we'll be talking in terms of voyages that will take anywhere from 5 to 30 years.

Such voyages don't seem very likely. It would be easy to take the pessimistic view and to suppose that manned exploration of the universe got off to a good start only because of the earth's good fortune in having a large nearby satellite and that that good start is also a sad end. Except for the moon and, perhaps, for the occasional minor wanderer into our vicinity, humanity may be in prison, kept in place by unbridgeable gulfs of emptiness. In that case, exploration in the classic sense has about reached its limits, and there is nothing left to do but fill in the details.

On the other hand, humanity is not necessarily confined to the earth and the moon, even if we don't venture beyond the moon. Scientists have speculated on the possibility of building space settlements in lunar orbit, with each settlement capable of holding from ten thousand to ten million settlers. They have concluded that to do this is quite within the capacity of our present technology, let alone one that is advanced to twenty-first century levels.

What difference will space settlements make?

Consider that the "unbridgeable gulfs" I mentioned are unbridgeable chiefly for psychological reasons. We can surely build machines that will endure in the relatively benign environment of space for years. (Our probes, untended by careful human hands while in progress, manage to survive.) We can also devise life-support systems that will sustain human beings for that length of time.

The chief difficulty will probably lie in the inability of human beings to endure confined quarters for so long a time or to adjust to an environment so different from anything they are accustomed to.

As soon as we imagine space settlements to be in existence, however, and human settlers to be living out their lives on them, the situation changes.

To the space settlers, dwelling in their small worlds in lunar orbit, space travel will be something familiar, not exotic as it is to earth-people. Even if the space settlement is as earthlike as it can be made, most of the settlers will be working part-time on the moon, on the construction of other objects in space, or on import-export through space.

Within the space settlement, the settler will be accustomed to a gravitationlike pull that changes in intensity from place to place, to living on the inside of a small world, to being part of a tightly cycled system of air, water, and food. To all these things earth-people are not accustomed and yet they are the precise characteristics of the space-travel environment.

Then, too, large space-liners can be assembled in space much more economically than on the earth's surface and, on taking off, will not have to escape the full pull of earth's gravity. The true spaceships of the future will surely be space-built by space-settlers.

What it all amounts to is that space-liners will not be very different from the space settlements themselves. The space-liners will be smaller and will hold fewer people, but they will be very much like home to the space-settler crew.

It would seem almost certain, then, that the space explorers of the future will not be earth-people, but space-settlers. It will be those settlers, emerging from their artificial homes in orbit about the earth, who will be the great navigators of the ocean of space—the Phoenicians, Polynesians, and Vikings of the future.

They will find no psychological obstacles to long journeys in space, and they will be the first to land on Mars and, perhaps, to build a permanent settlement there.

Beyond Mars will be the real bonanza of the solar system—the asteroid belt. There they will find 100,000 worlds with diameters of one kilometer and up. There will be metal asteroids, rocky ones, and icy ones, and each variety will have its own sort of usable resources. The asteroids will be the mines of the future, the industrial base for the civilization that will span the inner solar-system and lay the foundation for the still more magnificent ships that will carry man through the much vaster spaces of the outer solar-system.

And, eventually, the newer, larger settlements that will be circling the sun in the asteroid belt and beyond, carrying tens and hundreds of millions of human beings in societies that are in carefully designed ecological balance, may break away altogether. They may leave the solar system and set off on an endless trek through interstellar space, building their own lives for generations and millennia and, every once in a long while, passing new planetary systems within which they may renew their resources and upon which they may set up new habitations, buds and shoots of the old.

Most exciting of all, perhaps, they may encounter other forms of intelligent life.

If, then, we use the present reach of exploration properly, there need be no ultimate horizon for ages to come and humanity can expand virtually endlessly through the virtually illimitable universe.

# 53

# Homo Obsoletus?

I am constantly being asked to peer into the future in this direction or that, and frequently I am asked to consider the future of computers.

I am glad to do this and am quite capable of talking very rapidly on the subject, but sooner or later (usually sooner) I am interrupted in my course of glowing optimism and am asked, "But do you think that human beings may be replaced by the computer? That human beings may become obsolete?"

Do I? Let's consider the matter in orderly progression.

1. *Ought the question to be considered at all or is it just a very human fear and distrust of change, particularly technological change?*

One can imagine the anger, for instance, of early builders when the equivalent of the yardstick came into use.

One can almost hear them mutter, "Of what value then is the keen eye and the cool judgment of the experienced carpenter if any fool can tell whether a piece of timber will stretch across a doorway by measuring it with that inanimate marked stick? Brains will decay and the human being will become extinct, replaced by wood."

And surely the bards of old must have been horrified at the invention of writing, of a code of markings that eliminated the need for memory. A child of ten, having learned to read, could then recite the *Iliad,* though he had never seen it before, simply by following the markings. How the mind would degenerate!

A Spartan monarch, on seeing a demonstration of a catapult hurling its heavy rock, cried out, "Oh, Heracles, the valor of man is at an end."

He equated martial valor with hand-to-hand thumping, you see; but if so, he was too late by some thousands of years, for such a cry must have rung out with the invention of the bow and arrow.

These fears were wrong every time.

The use of inanimate aids to judgment and memory did not destroy judgment and memory. We use them all the better now for not wasting them in ways that a few marks on a piece of wood or on a piece of paper make unnecessary.

To be sure, it is not easy to find someone nowadays with a memory so trained that he can reel off long epic poems — but it wasn't so easy in ancient times, either, or a good bard would not have been as valued as he was.

And even if our unaided talents have degenerated a little, is the gain not worth the loss? Could the Taj Mahal or the Golden Gate Bridge have been built by eye? How many people would know the plays of Shakespeare or the novels of Tolstoy if we had to depend on finding someone who knew them by heart and was willing to recite them to us?

Then came the Industrial Revolution and its steam engines, and internal-combustion engines, and explosives, to take the weight of hard labor from the backs of human beings. Steam and gasoline vapor and electric current drag loads no horse could budge. Rocks are shattered in a moment that an army of slaves would take a week to split. Tricks with light and magnetism and subatomic particles are performed that no Scheherazade could imagine.

Do human muscles grow flaccid as a result? Yes, they might, except that they don't have to.

Keeping one's body in shape is the great game of today and people go through the motions of jogging and tennis and push-ups to make up voluntarily for what they no longer need do under the hard grip of enslaved compulsion.

And now we have computers. We even have very cheap, very small, very clever ones; computers that can do all the little tasks about our house and office that till now we had managed to do in our head, or with pen and paper — and which we so often got wrong.

We no longer need our ability to multiply eight and seven in our heads and get fifty-six (or is it fifty-four?), and we can discard our talent for making lists and forgetting the crucial item.

Will our brains therefore decay (for the thousandth time in the history of technological advance), and will we become obsolete?

Or will we once again make a virtue of a loss, and use computers as adversaries in games, for instance, that will hone our minds to new acuteness? Will computers do our work while letting us sleep in permanent stupor, or will they free us of disgraceful tasks beneath the level of human ability and allow us to tackle truly creative tasks — so that we may build Taj Mahals in place of mud huts?

It is, in fact, up to us whether to use our tools as cushions or spurs.

But wait, are computers tools? Just tools?

When I am asked whether computers will make human beings obsolete, the questioner does not have in mind a computer that serves simply as a tool, but one that serves as a surrogate-human.

After all, because a computer is a lump of inanimate matter notable only for the speed with which it performs simple arithmetical operations in endless and varied repetition, that does not mean it will stay so forever. At the rate at which computers are advancing and improving, might we not expect that, given enough time, computers will become capable of duplicating any feat of the human brain—any feat at all?

And, eventually, if computers can write books, devise poems, compose symphonies, perform research, create new ideas—will they not be as intelligent as human beings, or even more intelligent? And might they not then replace human beings, kill us off as unnecessary excrescences, and take over as the new masters of the earth?

If this is the case, we would have to decide that the human fear of the computer is an entirely new terror and a justified one, and not merely a repetition of a thousand fears of the past. The computer, it could be, differs from earlier technological advances in kind and not in degree only.

So we must consider the possible obsolescence of humanity and put to ourselves a second question—

2. *Why not?*

The history of the evolution of life is the history of the slow alteration of species, or the bodily replacement of one species by quite another, whenevei it happens that the change, or replacement, results in a better fit within a particular environmental niche.

In general, we human beings, as spectators of this past drama, tend to cheer on the victors. We think it only right that the vertebrates, possessing as they do an efficient internal skeleton, should now dominate the world of life, even though they are but the most recently developed of the grand divisions of the animal kingdom.

The conquest of the land is one of the great feats of the ages, and we approve as, first, amphibians, then reptiles, and finally the mammals, dominate the continents.

It is exciting to see brain power come into its own. The mammals are brainier than the reptiles they replaced; the placental mammals are brainier than the marsupials; the primates are the brainiest of all (if we ignore the cetaceans, who spoiled everything for themselves by returning to the sea and losing their chance at manipulative appendages).

Even in the past twenty million years we have seen the primates sort themselves out, the hominids finally appear on the scene, and, in climax, *Homo sapiens* come along to establish dominion over the world and to produce a technological civilization through the sheer force of a giant brain.

Of course part of the fascination of the drama is that we know and approve the ending. We view all the replacements as steps on the road to ourselves — and we are smugly satisfied that we ourselves are the crown and climax of the long trudge up the three-billion-year road.

Our pleasure in ourselves-as-climax makes any notion of a continuation of this play-that-is-over seem in the highest degree unnatural and reprehensible.

Yet the play is not over. It is only the accident of the briefness of our lives in relation to the speed of evolutionary change that makes the pattern of life seem static now; and it is the folly of self-love that leaves us satisfied with that. Actually, evolution continues and there is nothing unnatural in the thought that *Homo sapiens* will be replaced by a modification of itself slowly formed over the coming ages — or even by an entirely different species that better fits the environmental niche we now occupy (or the environmental niche into which our present one will change).

Of course the brain power of *Homo sapiens* and the accumulated machine-power of our technology is such that the simple evolutionary change of the past may no longer be in the cards. Human beings are now developing the capacity to engage in genetic engineering so that they may be able to guide their own evolution at a far greater speed than the blind force of hit-and-miss mutation and natural selection could manage in the past.

Human beings are also developing the capacity to create an artificial intelligence comparable to their own.

In either case, it may be that the grand design of evolution includes the slow change from species to species through random factors until, finally, after an extended period, a species is formed that is sufficiently intelligent to direct its own evolution and to create new kinds of intelligence on a nonorganic basis.

In that case, the replacement of humanity by either a hyper-humanity or by computers is a natural phenomenon to which we can object only for reasons that are frivolous and irrelevant.

So far, however, I have only been arguing that the replacement of humanity is not necessarily an evil. Can we go farther and say that it is positive good?

Perhaps.

Look about you and consider what human beings have done and are doing to the world. Consider the manner in which they have brought extinction to other life-forms, unbalanced the ecological relationships of those species that still remain, destroyed the soil, polluted the water and the air, introduced poisons and dangers the planet has never yet seen. Consider further that all these changes for the worse have been going on at an accelerating pace and are still accelerating now.

Viewed in that manner, the succession of a computer intelligence that is superior to the human variety and that is (perhaps) not associated with the emotions and with the judgmental incapacity of the latter is something that could be much to be desired. The great fear might be that the computer will not be developed to the point of succession before *Homo sapiens* succeeds in destroying itself and much of the planet as well.

It is with that thought in mind that sometimes, when I am asked if the computer will ever replace the human being, I answer, "I hope so."

But wait, are we sure that we will be replaced by a superior intelligence if and when one exists? Let us go on to our third question —

3. *What is a superior intelligence?*

It is entirely too simple to compare qualities as though we were measuring lengths with a ruler. Because we are used to one-dimensional comparisons and understand perfectly what we mean when we say that the distance from New York to San Francisco is greater than the distance from Chicago to San Francisco, we get into the habit of assuming that all things may be so unsubtly compared.

For instance, a zebra can reach a distant point sooner than a bee can, so that we are justified, we think, in saying that a zebra is faster than a bee. And yet a bee is far smaller than a zebra, and it flies through the air, which the zebra does not. Both differences are important in qualifying that "faster."

A bee can fly out of a ditch that holds the zebra helpless; it can fly through the bars of a cage which holds the zebra prisoner. Which is faster now?

If A surpasses B in one quality, B may surpass A in another quality. And, as conditions change, one quality or the other may assume the greater importance.

A human being in an airplane flies more quickly than a bird, but he or she cannot fly as slowly as a bird, and slowness may be very desirable for survival at times.

A human being in a helicopter can fly as slowly as a bird but he or she cannot fly as noiselessly as a bird, and silence may be very desirable for survival at times.

In short, survival requires a complex of characteristics, and no species is replaced by another because of a difference in one characteristic only. And that goes for a simplistic superiority in intelligence, too.

We see this in human affairs often enough. In the stress of an emergency, it is not necessarily the person with the highest IQ who wins out; it could be the one with the greatest resolution; the greatest strength; the greatest capacity for endurance; the greatest wealth; the greatest friendships in high places. Intelligence is important, yes; but it is not all-important.

For that matter, intelligence is not a simply defined quality; it comes in all varieties. The intensely trained and superscholarly professor who is as a child in all matters not pertaining to his specialty is a stereotypical figure of modern folklore. Nor would we be in the least surprised at the spectacle of a shrewd businessman who is intelligent enough to guide a billion-dollar organization with a sure touch and yet who is incapable of learning to speak grammatically.

How then do we compare human intelligence and computer intelligence and what do we mean by "superior" intelligence?

Already, if we wish to define intelligence as the capacity to perform arithmetical operations speedily, the computer is millions of times as intelligent as a human being—and yet we are all confident that the computer is not intelligent at all.

However, as computers are designed with greater and greater capacities, as they are designed to play chess, to translate languages, to compose music, to imitate the responses of a psychiatrist, it will become more and more difficult to maintain that it is not intelligent.

Remember, though, that the development of intelligence in human beings and in computers took different paths and was driven along by different mechanisms.

The human brain evolved by hit and miss, by random mutations, making use of subtle chemical changes, and by a forward drive powered by natural selection and the need to survive in a particular world of given qualities and dangers.

The computer brain evolved by deliberate design as the result of careful human thought, making use of subtle electrical changes, and by a forward drive powered by technological advance and the need to serve particular human requirements.

It would be very odd if, after taking two such divergent roads, brains and computers would ever end in such similarity to one another that one of them could be said to be "superior" in intelligence to the other.

It is much more likely that, even when the two are "equally" intelligent, the properties of intelligence would be so different in each that no simple comparison could be made. There would always be some activities to which computers would be better adapted and others to which the human brain would be better adapted; that this would be particularly true if genetic engineering makes it possible for human beings to improve the brain as well as the computer.

Indeed, it would be undesirable, perhaps, to try to develop either a computer or a brain to possess "all-around" capacities. The gain in generality would surely involve an inevitable loss in specialized abilities, so that keeping the two forms of intelligence different would remain desirable.

Consequently, the question of replacement is quite likely never to arise. What we would see, instead, would be a matter of complementation. It could be that human and computer might form a symbiotic intelligence that would be far greater than either could develop alone, a symbiotic intelligence that would open new horizons and make it possible to achieve new heights.

In fact, it could be the doorway that leads humanity from its isolated infancy to its in-combination adulthood.

## 54

# Volatiles for the Life on Luna

Now that the twentieth century has seen six space flights safely reach the moon and return, it seems natural to wonder if the twenty-first century will see the establishment of an ecologically independent colony on the moon.

The moon is in many ways an ideal site for the first extraterrestrial colony of mankind. It is only three days away by rocket flight, and radio communication can traverse the distance in 1¼ seconds.

To be sure, the moon has a day and a night that are each two weeks long, with temperatures rising to the boiling point of water in some places at some times, while dropping to sub-Antarctic chill at other places at other times. There is also the hard radiation from the sun and the potentiality of meteor strikes, since there is no atmosphere to ward off either.

These are surface manifestations only, however, and if the colony is established in carefully engineered caverns beneath the surface, there would be equable temperatures at all times and excellent security as well.

There would remain the problem of lunar gravity which is only one-sixth what it is on the earth's surface, but it isn't unreasonable to hope that people can adapt to this.

As to the amenities of life, the problem of energy supply should not be a troublesome one. By the twenty-first century, fusion power should be available; and, even if it were not, there is solar energy. There are millions of empty square miles on the moon's surface and no native ecology to be disturbed. The sun, shining down unbrokenly for two weeks at a time upon large arrays of solar batteries, should supply all the energy the colony needs. The proper placement of as few as three such arrays would ensure energy production at all times, since one or another would always be exposed to sunshine.

Of course there is no atmosphere on the moon and no bodies of standing or running water, but if only the soil of the moon were basically like that of the earth, this would not represent an insurmountable problem.

Water could then be obtained from the rocks themselves (as could metals and other important materials, given an ample energy supply). The water could be electrolyzed to hydrogen and oxygen. The oxygen could be used for building an atmosphere; the hydrogen as a source of fusion fuel, or in chemical syntheses. Algae, growing under artificial light, and fertilized by properly treated human waste (plus chemical fertilizers from lunar soil) could renew the atmosphere and serve as a food supply.

Eventually some forms of animal life could be introduced, to say nothing of plant life less efficient than algae in converting carbon dioxide and water into food. A reasonably normal human dietary might be established.

Since the colonies would carefully recycle everything as efficiently as they could (as otherwise survival would be questionable), not much in the way of new material would have to be introduced. Small capital investments of water, for instance, would last a long time. Additional supplies would be required more to sustain colony growth, perhaps, than to replace cycling losses.

This picture of a viable lunar colony, however, breaks down in the light of the study of the lunar rocks brought back by the astronauts. Rather disappointingly, the moon's crust seems to be low in the content of the more volatile elements — those with compounds that are low-melting. Presumably, the moon went through more or less extended periods at elevated temperature and lost them by vaporization.

In particular, water is absent. Judging by the nature of the lunar rocks we have studied, it would seem that the lunar crust is everywhere bone-dry.

If this is so, does it completely eliminate the possibility of a lunar colony? (Can we expect society to develop in a total water-desert?)

The problem of water-lack adds a complication, to be sure. Volatile materials, and especially water, must be obtained from somewhere other than the moon if a lunar colony is to remain possible. The logical source for the volatiles is, of course, Earth itself. The colonizing expeditions can bring water with them, and further supplies can be sent periodically and, by recycling, made to last as long as possible.

This may not be as bad as it sounds. After all, water can be sent to the moon more cheaply than people can. A water cargo does not require a ship with expensive and complex life-support systems. It may be that water will be lifted to a space-station by rocket-shuttle. There it can be frozen by a method as simple, perhaps, as allowing part of it to evaporate when exposed to the vacuum of space. Large slabs of bare ice might then be fired off to the moon by some advanced technology that would allow part of it to be vaporized and made to serve as a rocket exhaust calculated to send the rest to the moon in a slow trajectory.

There would be no lunar atmosphere to vaporize the ice at its approach to the moon, and loss to solar radiation may be small. Undoubtedly techniques will have been developed to minimize melting and also to "field" the slab of ice and bring it safely to the lunar surface once it has reached the immediate neighborhood of the moon.

Nor need Earth feel it is giving up an indispensable natural resource. Earth has no shortage of water. When we do speak of water shortages, we are speaking of fresh, unpolluted, liquid water — which makes up only a small fraction of Earth's water supply, and which need not be touched.

There is sea-water, which makes up 98 percent of all the water on Earth and which we can well afford to give up in small quantities. Frozen sea-water can easily be distilled on the moon, given indefinite quantities of solar energy, and both the water and the salt-content would be useful.

But suppose this simplest of volatiles for lunar colonists — Earth — fails them. Perhaps the difficulties of lifting large quantities of water to the escape velocity and of getting it through our atmosphere to a space-station, of then freezing it and hurling it with pin-point accuracy to the moon, may all become too expensive to please Earthmen.

It may be politically objectionable to devote so much effort for the sake of lunar colonists. It may be psychologically objectionable to give up water, which all men have been taught to look on as a basic essential for life, even when it can be argued that it can be spared.

For that matter, the lunar colonists themselves may be unenthusiastic about receiving water from earth — feeling that it ties them to the mother world and prevents them from experiencing a true independence.

Is there any other place besides Earth, then, from which the colonists can get the necessary volatiles?

Of the objects that are permanent members of the inner solar system, only two, other than Earth, have a reasonable supply of volatile matter. These two are Venus and Mars. On Venus, which is extraordinarily hot, the volatile material is in gaseous form and therefore difficult to gather. The gravitational force of Venus is almost equal to that of Earth, so that what is gathered is hard to drag away. On the whole, it is difficult to imagine Venus as a practical source of volatiles.

Mars is much better in some ways. Its temperature is low and it has sizable icecaps containing frozen water and frozen carbon dioxide, both desirable materials for the lunar colony. The Martian gravity is only two-fifths that of Earth's and its atmosphere is very thin. This means the conveniently condensed material of the icecaps can be lifted off the Martian surface and through the atmosphere with far less trouble than would be true on Earth, for instance.

The catch is that Mars is never closer than 56 million kilometers to the moon. To get there, secure the ice, and bring that ice back would be a formidable undertaking. A permanent station might be established on Mars, perhaps, the function of which would be to fire ice into space in an orbit intersecting the moon's position (a much more delicate bit of aiming than would be required from an Earth station).

Yet if man's space technological expertise becomes such that the retrieving of volatiles from Mars is practical, it would have also become practical to establish a colony on Mars. In some ways, Mars would be a more comfortable home for man, since it would have a higher gravity than the moon, have some atmospheric protection against meteors and against weaker radiation, and, most important, have a supply of volatiles. In that case, Martian volatiles would surely be reserved for Martian colonists.

Where else can the moon turn? It is in the outer solar system that the real supply of volatiles is to be found. Giant Jupiter may be made almost entirely of volatiles, and its larger satellites have an ample supply, too. Callisto, the large satellite farthest from Jupiter, would seem to be particularly rich in volatiles from its low density.

Even so, Callisto is never closer than 730 million kilometers to the moon (thirteen times the nearest distance of Mars) and, although nearly two million kilometers from Jupiter, is still uncomfortably close to that planet's huge gravitational field. What's more it is within Jupiter's fierce radiation belts. And other possible worlds are farther away still.

No, it is not likely that the outer solar system will be within reach in the crucial decades when the lunar colony is fighting for life.

But not everything that exists in the outer solar system remains there permanently. There are some objects in the solar system that do *not* have orbits that are nearly circular, as the planets do, but have orbits, instead, that are enormously elongated. Consider the comets. At the far end of their orbits they are in the outer solar system, far beyond even the farthest planet in some cases. At the other end of their orbits they pass through the inner solar system.

The comets, originating (it is thought) in the far reaches of space a light-year or more from the sun, and consisting of remnants of the cloud of dust and gas from which the solar system originally formed, are composed largely of volatiles. They are made up of frozen compounds of carbon, hydrogen, oxygen, and nitrogen (the very elements that make up 99 percent of living tissue) with an admixture of rocky materials.

Comets lose some of their volatiles at each approach to the sun, those volatiles boiling off to form a foggy "coma," which is then driven outward by the solar wind into a long, filmy tail. Those comets which have been influenced by planetary gravitational fields into taking up a short orbit that brings them to the neighborhood of the sun every hundred years or less have lost much or most of their volatiles.

Every once in a while, though, a comet from the far-out belt is maneuvered into an elongated orbit by the tiny gravitational influences of the nearby stars. It can come streaking into the inner solar system for the first time and will then have all its original supply of volatiles on board, so to speak.

It may be that before the twenty-first century is over, the lunar colonists, having struggled along with what skimpy supplies of volatiles they have squeezed out of reluctant Earth, will have developed the techniques for trapping such comets.

Lunar-based telescopes would detect such comets far out in space even before they reach Jupiter's orbit. (To be spotted far off is itself the sign of a large new comet.) The comet's orbit can be plotted and, in the months it takes to pass into and through the inner solar system, the lunar colonists will have placed a ship at some rendezvous point in space.

A landing will be made on the comet (which may be no more than a few kilometers across the solid core). Rockets appropriately placed upon it (or, very likely, the use of some advanced nuclear drive) would force the comet out of its orbit.

Little by little, the comet's motion will curve in such a way as to bring it slowly closer to the moon and then into orbit about the moon, and then spiraling down to the moon's surface. Finally, it can be brought down within the southern lip of a north-polar crater, for instance, where in the eternal shadow of the crater wall it will remain permanently frozen.

The whole process will be like that of hooking, maneuvering, and landing a gigantic fish. And the "beached" comet will bring prosperity to the moon, as a beached whale brings a food supply to an entire Eskimo village.

Such a comet, with cubic kilometers of volatiles, will easily make the lunar colonists independent of further supplies for decades; for a century, perhaps. And by then another comet may have heaved into view.

The important task of the lunar colonist of the twenty-first century, then—the most exciting sport, the most difficult art, the most glamorous accomplishment—may be something as yet unimaginable in its details. It will be the sport and business of comet-fishing.

# 55

# Touring the Moon

It is the year 2082 and the moon is a settled world. There are 50,000 people who consider themselves Lunarians and who accept the moon as their home, and of these more than 5,000 have been born here and have never visited Earth. When tourism is at its height the total population is well in excess of 100,000.

Lunarians view tourists with mixed emotions. On the one hand, tourists crowd the space lanes and, at times, overload the moon's living facilities — and the moon, despite all the advances of the past century, is still not an open world. Its available air and water must be carefully recycled and every drop of water replacement (hundreds of thousands of gallons per year) must be imported.

On the other hand, the Lunarians are proud of their world, have an almost feverish desire to counter the stereotype of the moon as a bleak and desolate place, and (let us admit) can make use of the money that tourists bring.

Most tourists who arrive are first-timers, people who have never left Earth before. They arrive after a three-day journey in which they have experienced the thrills and inconveniences of weightlessness, and look forward with relief to reaching the surface of a world where up is up and down is down. Despite all indoctrination, however, they seem to expect only one kind of world, one with Earth's surface gravity.

This misconception is heightened by the fact that every effort is made to give the moon an Earth-like appearance. The ship does not land on the surface, which is undeniably *bleak* (though Lunarians never use the word and would prefer to banish it from the dictionaries). The ship sinks into a huge airlock, and the passengers eventually step into a large visitor's entrance port, in which atmosphere, temperature, and decor are completely Earth-like. What cannot be changed, however, is the moon's surface gravity, which is one-sixth that of Earth.

Nothing, apparently, can prevent that from being a surprise to first-timers. After the initial shock, the reaction is inevitably amusement, and a tendency to try walking, hopping, or jumping, despite the large signs that ring every possible change on the message, "Please do not run or jump, but wait quietly for processing."

This exasperates Lunarian officials, who have difficulty in maintaining order, and who are particularly disturbed by the occasional falls. Still, the low gravity usually prevents any damage, and falls but add to the hilarity.

The first day is usually a particularly tedious one, for every visitor to the moon must be thoroughly examined biologically and medically, despite the initial screening on Earth. No undesirable life of any sort — seeds, parasites, germs — are permitted in a world in which the ecological balance is carefully controlled and maintained.

And the first night, by all accounts, is invariably a restless one as natural night movements tend to heave one upward unexpectedly. First-timers quickly understand the reason for the padded barriers running along every edge of a Lunarian bed.

By the second day, most tourists are acclimated to the low gravity and are willing to venture out and begin exploring the moon. For that purpose there are the lunar vehicles, which are so characteristic of the moon that their stylized representation serves as the universally recognized symbol of our satellite as an inhabited world. In these sturdy and maneuverable rocket-powered vehicles, passengers would remain perfectly safe if they wore ordinary clothing, but regulations require them to wear spacesuits.

To be sure, these spacesuits are not the bulky and cumbersome affairs of the early astronauts (which are what the word, spacesuit, seems to imply to Earth-people even today) but are remarkably little different from ordinary winter clothing, except for their impermeability, the oxygen cylinders discreetly attached, and the arrangement whereby a helmet can be clicked tightly into place in one movement. Under ordinary conditions, the helmet is suspended from the chest — clumsy, but necessary under the rules.

There are two major areas on the moon's surface that are musts for first-timers and these do not include any of the natural formations. There is a certain austere interest in the mountains and craters of the moon, but there is no denying that old bromide: "See one lunar crater, and you've seen them all." Tycho on this side of the moon and Tsiolkovsky on the far side get their share of tourist attention, but there is frequently expressed disappointment. The fact is that Earth's mountains are more rugged, and the icecaps on land and the complex life patterns undersea lend them a grandeur and interest the moon cannot duplicate.

Not so the two major human additions to the lunar landscape. First is the great lunar mining complex at the Neil Armstrong rift. Almost every step in the mining process is automated and in charge of robots. The tourists view it at night, of course, since it is not really practical to remain bathed in the heat and hard radiation of the sun for any length of time, considering that there is no natural atmosphere to serve as protection.

Solar energy on the moon is cheap, however, and the mining complex is well lit. The ship lands on the lip of the rift and the tourists affix their helmets (each one of which is carefully examined by the stewards on board) and emerge to look down at the absolutely unbelievable panorama. There is the vast pit that has been dug out by the activity of seven decades, but within which there remains an almost unimaginable further supply of metal-rich ore. A never-ending chain of buckets move along rails to the mass driver, where an electromagnetic field accelerates them and flings them out into space like the world's largest slingshot.

Almost all the structures, from factories to settlements, that now dot various portions of "cis-lunar space" (the region between Earth and the moon) have been built of material obtained originally from this gigantic hole on the moon's surface.

The second sight, smaller but more intensely human, is the great Karl Jansky radio telescope on the far side of the moon. This is also best viewed at night; and, considering that day and night each last two weeks on the moon, a tourist cannot see both the mining complex and the radio telescope without usually having to wait anywhere from two to ten days in between.

The radio telescope dwarfs anything of the sort on Earth or in space. It is a kilometer in diameter and, with its auxiliary equipment placed elsewhere on the far side, its effective diameter is virtually that of the moon itself. This radio telescope has the full width of the moon between itself and Earth and is free of all but minor stray radio interference from space establishments. It has, in recent decades, detected several radio-wave patterns from fairly nearby stars that may indicate the presence of extraterrestrial intelligence. (Astronomers are still arguing.) Tourists, led through the underground laboratories, watch with rapt attention as the needles mark out the delicate rise and fall of those microwave intensities that may indicate nonhuman intelligence.

Touring the underground micro-life vats on the moon interests many, since nutrients are produced in great quantities here. Not only does this help feed the Lunarians, but increasingly, they serve as food additives for Earth-people. It must be admitted, however, that the odors encountered in the gloomy chambers are not to the ordinary taste.

The Lunarian shows are famous, and they are not for export since the low surface gravity is of the essence. Expert Lunarian gymnasts can perform "gravity-defying" feats that are simply impossible on Earth, even for gibbons. Tourists cannot fail to be enthralled.

It would seem at first glance that skiing on the moon is impossible. For one thing, where is the snow? Snow, as it turns out, is not needed. Under the low gravity, the body presses down against the gritty lunar surface only mildly and this reduces the friction to a large extent, making that surface surprisingly slippery. Add to this the manner in which professionals attach small cylinders of argon gas to their shins. This produces a layer of gas under the boots that further decreases friction. Down the gentle slope of a lunar crater and across the lunar surface (which invariably supplies sufficient unevenness to make an excellent obstacle course) the skiers race with incredible grace.

Invariably the more athletically inclined of the tourists try their hand at it themselves and, though they are helped along, given easy equipment and gentle slopes, they just as invariably find that it is not as easy as it looks. Again, fortunately, the inevitable falls are not as hurtful as they would be on Earth.

There is no question, though, that of all the tourist attractions on the moon the most absorbing is the sky. In some ways, the lunar sky does not differ significantly from that of Earth. Earth and the moon occupy the same region of space and we see from the moon the same stars, arranged in the same constellations, that we see from Earth.

The moon, however, has no atmosphere, and therefore no fogs, mists, smoke, or clouds to interfere with an always perfect view. From behind the high-transparency glass of an observatory, every star (non-twinkling) is about 25 percent brighter than it would seem from Earth. The planets, too, are brighter than we are accustomed to, and Venus at its brightest is almost mesmerizing in its brilliance.

The sky moves at only one-thirtieth the apparent speed of Earth's sky, for the moon is a slowly rotating object. In a way, this is a disadvantage, for one can get tired of an almost unchanging view. On the other hand, it allows a fourteen-day period of night before the sun appears on the eastern horizon. Once the sun appears, of course, direct viewing becomes impossible. People must then watch the sky indirectly by means of a computerized-television set-up in which the sun is selectively screened out.

The one important object we don't see in the lunar sky is, of course, the moon. In place of it, however, is Earth. Earth is an excellent substitute, for it goes through all the phases of the moon in the same order and the same time, but Earth is about 13 times larger in surface area in the moon's sky than the moon is in ours. What's more, Earth reflects more light so that, when full, it is 70 times as bright as the full moon is in our sky. As the Earth-phase narrows, the discrepancy is even larger in Earth's favor. Furthermore, Earth is covered with a swirling cloud pattern that is always changing and is endlessly fascinating to watch.

Because the moon always keeps one side facing Earth, Earth hangs in the moon's sky forever, without changing place (at least when viewed from the side facing Earth). It does move in a slow, small ellipse because of a lunar movement called "libration," but this is scarcely noticeable and is never remarked on by the tourist

Eventually, the sun appears in the east, and direct viewing ends. On television one can see the sun cross the sky, in a two-week journey, and pass Earth either above or below. As the sun nears, Earth becomes a thinner and thinner crescent and then, after the sun has passed, the crescent widens again.

Every once in a while, the sun passes behind Earth so that its rays are blocked and do not reach the moon. On Earth, we see this as an eclipse of the moon.

This is the supreme sight on the moon. In the midst of the two-week-long day, night falls and can endure for up to two hours. The television is turned off and direct viewing is possible again.

The stars are at their most brilliant and Earth is by no means invisible even though the sun is behind it, so that we see our world drenched in its own night. What saves the situation is that the sunlight strikes the atmosphere on every side of Earth. The short-wave light of the sun is scattered, but the long-wave red and orange light passes through and reaches the moon.

Earth then becomes a black circle in the sky, within which no star can be seen, a circle that is rimmed by a thin edge of brilliant red-orange light. The circle is nearly four times as wide as the full moon appears in Earth's sky. Depending on where the sun is behind Earth, the circle of red-orange light will be brighter on one side than on the other. When the sun is centered behind Earth, the circle is equally bright all around. There are bound to be clouds along the rim of Earth, so that the circle may be broken here and there and, at particularly unlucky times, hardly any of it may be visible. At particularly lucky times, all of it may be visible.

In any case, an eclipse of the sun, as seen from the moon, produces a vision of a kind that can never be seen from Earth. The beauty of it must be experienced, for descriptions are never adequate.

It is not surprising, then, that tourism is always heavy at times when an eclipse is scheduled. Astronomers can predict, far in advance, when these times will come, but there is not much use in urging my readers to make their reservations now. All shipping space to the moon at eclipse time is booked twenty years in advance.

---

# —56—

# Life on a Space Settlement

The thing to remember is that a well-designed space settlement is a whole town, even a city, but to its inhabitants it will be a world.

When you think of a space settlement, there's no use in thinking of Skylab. That would be like thinking of Columbus's *Santa Maria* when you should be thinking of *Queen Elizabeth II*.

There are objectives to be gained in nearby space. There is the collection of solar energy by power stations in orbit about the earth, advanced satellites of all kinds, astronomical observatories, laboratories, automated and computerized factories—all in orbit about the earth. There will be mining stations on the moon that will supply the metals, concrete, glass, and soil needed for all these structures.

Very little of all this, however, can be accomplished by human beings based on Earth. The work will be too difficult and expensive if those performing it must commute from Earth's surface. The reason why space settlements will be built will be to serve as bases for the miners, the engineers, the construction workers, the scientists, and for their families as well.

To house them we will have structures half a mile or more across, in the shape of cylinders or spheres or doughnuts. The people will live in the interiors.

The prototypes of such settlements will be built with resources from Earth; but, once we have a start, the space settlers themselves will build more space settlements out of moon material. There will then be a rapid expansion as hundreds of space settlements, including quite large ones, are built.

Constructed of metal and glass, the space settlements will be coated with soil on the inside, thick enough to protect against cosmic-ray particles.

317

The settlements will be lit by mirrors that accompany them in orbit and that reflect sunlight through the windows, which, by a louver arrangement, can change the angle of light and periodically close it off altogether in order to mimic the day-and-night alternation to which human beings are accustomed.

The settlements will be set to spinning, to produce a centrifugal effect that will press everything against the internal surface of the settlements. It will seem to people that there is gravitation holding them against that internal surface, and the rate of spin will be chosen to make that feeling of gravitation a normal one as on Earth.

The interior can be built up to suit the tastes of the settlers. There can be farms, houses, streams, trees, churches — all the paraphernalia of American small-town life — if that is desired.

Not everything will be as on Earth, of course. In some ways conditions will be much better on the settlement. The ecology can be designed from the start with undesirable components omitted. There needn't be any poisonous snakes or noxious insects — if they can be kept out. The temperature can be equable, the weather mild, with floods, droughts, storms, frosts, and heat waves unheard of. Air, food, and water will be carefully recycled and waste will be minimized.

Conditions can be modified to suit needs. Different sections of a settlement can be semi-independent. In a section given over to agriculture, the period of light may be longer than usual, and the atmospheric content of carbon dioxide higher than usual in order to promote plant life. There may be a section of continuous night for nightclubs, open-air movie houses, and so on.

The first space settlements are likely to be American. The settlers will be American with American ways of thought and life, and the settlement will reflect that. No doubt the babies born on such settlements will be considered American, the settlers will pay taxes to the American government, and their work will be strongly subsidized by the American government.

This, however, would be a temporary situation. There are quite likely to be Soviet settlements following, and then perhaps those of other nations. In addition, the requirements for efficient work in space — the skills, the physical dexterity, the scientific knowledge — will not be terribly easy to meet and people from any nation will be welcome if they have what is needed. The settlements are sure to become cosmopolitan.

Add to that the fact that power stations in space will become the chief source of Earth's energy, that factories in space will become major suppliers of high-technology products of all kinds, that observatories and laboratories in space will become the great originators of new knowledge and techniques for use by humanity. All this is bound to give rise to a feeling of planetary unity on Earth. All the nations will be dependent on space and all will want to be involved in it. The earth itself will become more cosmopolitan.

What's more, the intricate technological web that will support the human thrust into space will be easily disrupted in case of major disorders on Earth, and the entire benefit of space development will be lost. So important will that development be that war on Earth will become unthinkable out of sheer self-interest.

How will the settlers adjust to their little worlds?

Worlds they may be, but they will be small compared to Earth. A good sized settlement may have a population of no more than 100,000 and any settler can walk about this world without too much effort and explore its entire surface. There will be nothing hidden, no more nooks to discover. Will there not be boredom? Even claustrophobia?

Probably not. The settlements will tend to be in the moon's orbit; some clustered before the moon, some behind. In moving from one to another there will be very little need to fight Earth's gravity. The moon's gravity will interfere somewhat, but its gravitational field is much less intense than Earth's.

People from one settlement can reach the next with almost no rocket power. It will be rather like gliding along an icy surface.

The common tasks of building structures in space, of servicing the various structures already built, of supervising the automated devices, and of reprogramming the computers will make all the space settlers used to rocket travel. Small space-skimmers, low-powered, easily maneuverable will be to the settlers what taxis are to New Yorkers or gondolas to Venetians. They will be a bonding agent and a means of travel so common as never to be given a second thought.

Indeed, increasingly the settlers would not think of themselves as belonging to one settlement alone. One particular settlement might be where the sleeping quarters of a particular family might be, but all the settlements together would be home.

In a world without war (which we may assume, or there will be no true space age) there may be rivalries and competition between settlements, but no deadly antagonisms. In fact, a common life and work is quite likely to produce a federation of space, a sort of "United Settlements" organization, which would supply a common citizenship for the settlers and be supported by taxes on them.

The fact that the various settlements might differ in culture (particularly in language) would make them the more interesting to each other — rather like the various ethnic neighborhoods of an American metropolis.

For that matter, it is possible that there will arise a common understanding — a common patois, for instance, built up of English and Russian, together with scraps of other languages. Most settlers would be able to make themselves understood on any settlement.

Because of the constant travel from one settlement to another, the settlements might develop a common ecology, no matter how different they might have been to begin with. In fact, the great differences and difficulties would be between the settlements generally, on the one hand, and Earth on the other.

Visits from Earth-people would surely be discouraged. Earth-people would, after all, be very likely to carry parasitic life-forms not wanted on the settlements. They might carry seeds and spores in their clothing.

Once a settlement is mature, its population would have to be carefully controlled, so that no immigrants would be allowed. Population expansion would come about only through the establishment of new settlements, and there it would be advisable (perhaps even compulsory) to accept a certain number of Earth-people. These would have to be quarantined and examined in detail for health and for parasites before they could qualify as full-scale immigrants planning to live out their future lives on the settlements.

In fact, the chief duty of the United Settlements organization would be to regulate such immigration and to establish and enforce the overseeing of trade inspections.

To be sure, there would be Earth-people interested in tourism, but they would have special settlements set aside for them where they could experience the amenities of settlement life. These settlements would be run by and for Earth-people only, and the true settlers would avoid them.

Nor would travel go in the other direction either. The settlers are not likely to want to visit Earth either on business or as tourists. Not only would they fear picking up infections unknown in the settlements, but they would not be eager to experience Earth's temperature and weather extremes, or to undergo the prolonged quarantines and treatments before they could return home. The chances are they would also find Earth psychologically uncomfortable because it would be a place where people lived on the *outside* of a world rather than on the inside.

To be sure, despite the difficulties of physical intercourse, Earth and the settlements would need each other. The settlers would make the products of space technology (including solar energy) available to Earth, while Earth would supply the settlers with those key elements — carbon, hydrogen, and nitrogen — that are unavailable on the moon.

And, of course, it is only physically that the settlers and Earth-people would be apart. There would be complete information-contact by holographic laser beam. Any settler could see any of the impressive sights on Earth, from the Grand Canyon to the Taj Mahal, to a rain forest, to a sand desert in the quiet of his settlement home. Earth-people could similarly visit the different settlements.

When relaxing in his home, a settler will find the sights quite different from anything Earth-people would see. Since the settler lives on the inside of a small world, whatever its shape, he would see no horizon. Rather the ground would curve upward quite quickly to his sight, unless the settlement were specifically designed to break up long views.

Indeed, if the settlement were properly shaped and small enough, a settler would see the other side of the inside-out world directly overhead. He could walk there without trouble, of course, for he would climb no hill. The small world would turn him so that he would always seem right-side up. Then, when he was on the other side, he would see his own house upside-down overhead.

It would not seem strange or frightening to him; he would be used to it. Rather, it would be on Earth that he would be frightened, when he saw the ground level come to a visible end at the horizon.

What forms would leisure take on a space settlement? After all, there would have to be leisure; settlers can't always be constructing, maintaining, and supervising. There are times off.

In many ways, settlers would spend their leisure time as Earth-people would. They might garden, picnic with their families, watch television, play cards or chess, indulge in parties, conversation, or sex.

All space settlements, however, have one characteristic Earth does not possess, something that would enormously affect leisure activity. Space settlements, of whatever shape or design, will have gravitational effects that vary in intensity from place to place; from Earth-normal, in places where the settlers go about their ordinary tasks, to lower and lower values in other places—all the way down to zero gravity.

Traveling from one part of the settlement to another may often involve passage through falls and rises in gravitational effect. This would have to be allowed for. Thus an elevator would be strongly pushed to one side as it rose, and to the other side as it fell, by something called the Coriolis force. These pressures would have to be allowed for in the design, and settlers would be acclimated to the feeling. Earth-people, when subjected to this on their tourist settlements, would find it as difficult to get used to this as land-lubbers would to the pitching of a small vessel at sea.

On the other hand, changes in gravitational effect would have their uses. Mountain-climbing on a settlement (on the larger ones, mountains a mile or two in height could easily exist) would be a delightful exercise. There would be neither snow, nor cold, nor thin air, and the higher you went the weaker the gravitational effect is likely to be, so the easier it would be to go still higher. Of course the element of danger would be gone to a large extent, but most people wouldn't mind that.

Ballgames of all sorts would require new skills if played under lower gravity. Balls would arc higher, come down more slowly; on the other hand, so would players. There would be a slow-motion grace to tennis, for instance, and a longer period of suspense while you wait for a slowly rising racket to meet a slowly rising ball over larger playing fields.

At zero gravity, games would gain a total three-dimensionality. There would be air-hockey, in which the goals would be six in number at the ends of an invisible octahedron (or eight at the ends of an invisible cube). Players would have to swim through the air in pursuit of a ball. Each player might be outfitted with "keels" along his back and abdomen to give stability and keep him, or her, from tumbling when he, or she, tried to move. There would be "fins" on arms and legs to make air-swimming more efficient. No doubt it wouldn't be easy.

The same with dancing, particularly ballet, under these conditions. Or trapeze work. Or flying for the fun of flying. Or calisthenics. Or playing tag. Or just horsing around.

All this is bound to be a wonderful sensation, beloved by all settlers. There would be areas reserved in every settlement for people who want to play or fly at zero or near-zero gravity. Children would probably learn to fly as early as they learn to walk. (Children may try to fly or walk under inappropriate gravity conditions at first and be frustrated, but they will learn.)

Low-gravity swimming and diving into water would also have its special fun and skill. And, for that matter, so would low-gravity sexual activity.

At all times, to be sure, there would be the danger of misjudging the true effect of weightlessness. While weight might be little or zero, mass and inertia would not change. A tumble from a height, or an unguarded collision with a wall, can bruise, cut, or even break a bone, regardless of how light we feel. Settlers will learn that early in life. Tourists from Earth on their tourist-settlements would have to be carefully indoctrinated and even so there will be occasional accidents.

The low-gravity activity would, more than anything else, distinguish life on a settlement from life on the constant-gravity environment of Earth. And that low-gravity activity would have its uses. In addition to being fun and adding to the interest and joy of life, it would keep bodies fit even in the soft, high-technology life of the settlements.

What's more, since so much of the serious work of the settlers would be on the low-gravity surface of the moon, or under zero-gravity conditions in space, the low-gravity play would prepare the body for low-gravity work.

One more thing. Would the settlers have goals that Earth-people don't?

Yes. To Earthmen, getting off the planet and into space would be a hard task, both physically and psychologically. To settlers, it would be easy; space-travel would be a "natural." To the settlers, there would be the goal of moving out to the asteroid belt where there would be more room for settlements and more easily available material for building them.

There would be the desire to explore the vastness of the outer solar system on long trips that the settlers could endure and Earth-people could not.

There would be the dream of some day sending entire settlements into space to go adventuring for indefinite periods toward the distant stars.

The settlements and the settlers would be the cutting edge of humanity. The settlers would be the pioneers. Human history would move outward with them and Earth's role would shrink. But then, Earth would, after all, have finally fulfilled its role of giving birth to a new and greater stage on which the human drama could play itself out.

---57---

# The Payoff in Space

It isn't easy to tell exactly what item uncovered in the ivory tower of science will be praised joyously a century later.

In 1677, a Dutch scientist, Anton van Leeuwenhoek, was the first to discover the world of microscopic life. It seemed of no importance in any practical way. If Congress had then existed and had given van Leeuwenhoek a modest appropriation to continue his observations, Senator Proxmire would probably have given it his Golden Fleece award. Nevertheless, two centuries later, the germ theory of disease arose out of findings concerning microscopic life and, as a result, the average life-span of humanity (and Congressmen, too) was doubled.

Nowadays, it is space exploration that endures the "So what?" attitude, the "What's in it for me?" question.

We might speak of the mountains on the far side of the moon, the craters on Mercury, the plateaus of Venus, the dead volcanoes of Mars and the live ones of Io, the global glacier on Europa—all revealed in the last couple of decades—and yet one can easily dismiss them all if one is sufficiently "hard-headed."

It is less easy to dismiss the weather on Earth, which affects us all, intimately and daily. How nice it would be to be able to predict it accurately. How much better to control it and make it do our bidding.

The trouble is that Earth's atmosphere is a very complicated machine—unevenly heated by the sun, swirled by Earth's rapid rotation, loaded with water vapor unevenly and unpredictably by Earth's oceans. We just can't get a good handle on it.

If we had simpler atmospheres to study and analyze, we might work our way up to Earth's complications. Well, now we do!

The atmosphere of Venus is uniform in temperature and there is no ocean on Venus to complicate matters. What's more, Venus rotates very slowly and doesn't give its atmosphere that added twist. However, Venus's atmosphere is very dense and that may complicate matters somewhat.

The atmosphere of Mars, on the other hand, is very thin and also circulates over a world that has no ocean. Mars, to be sure, rotates and is unevenly heated, though the effect is less important in each case than is true of Earth.

Jupiter's atmosphere is even thicker than Venus's, and it covers a planet that may be entirely liquid under the layer of gas. Jupiter's atmosphere is evenly heated over the spread of the world, though it gets steadily hotter with depth.

Each of these atmospheres may be, in its own way, a simpler machine than is that of Earth. If we can learn enough about atmospheric circulation on Venus, Mars, and Jupiter, then it may be that we can build on that knowledge to gain a better understanding of our own atmosphere—with incalculable benefits to mankind.

Though planetary exploration may seem to be pure ivory-tower research without practical importance, those who believe so demonstrate a failure of vision and imagination. The possible improvement in our understanding of weather is but one example of many that we could choose.

Enormous advances may arise out of space closer to home. We need go no farther than the moon to see that.

We need energy, for instance, and the problem is growing acute. Oil is dwindling, coal and nuclear fission are each perceived of, by many, as dangerous. Nuclear fusion has not yet been attained. Other sources are insufficient. What do we do?

There seems to be a growing consensus that solar energy is the way to go. If so, however, then the use of rooftops for heating and air-conditioning individual homes may be useful in conserving minor quantities of fuel, but it would supply only a small fraction of the energy needed by a working industrial civilization.

What we will need are solar cells that convert sunlight directly into electricity, and it may be necessary to coat tens of thousands of square miles of sunny desert areas with such cells. However, even the best areas on Earth are subject to the fact that it is night half the time and that even when the sun is shining and the air is cloudless and clear, much of the sunlight is absorbed by the atmosphere, especially when the sun is low in the sky.

If solar energy is absorbed by solar cells organized in a power station in orbit about the earth—free of atmospheric absorption, and with the incidence of night reduced from 50 percent of the time to 2 percent—each cell would produce up to 60 times the electricity it would on Earth's surface.

Electricity in space could be turned into microwaves and beamed down to Earth's surface nightly. It could there be picked up by relatively small receiving stations and re-converted to electricity.

Nor would space be a source of energy alone. It can become the great home of much of Earth's industry. In addition to the solar-energy stations, we can picture many automated factories in space. Such factories could make use of the special properties of space—hard vacuum, zero gravity, extremely high or low temperatures, hard radiation.

Each of these conditions could contribute to the production of alloys, machine components, electronic devices, pharmaceuticals, etc., of types and in ways that would be difficult, or even impossible, to produce on Earth's surface.

In addition there is an enormous volume of space into which to discharge wastes and pollution, diluting it all to harmlessness and even insignificance. Nor is it just a matter of space in Earth's immediate vicinity, for the solar wind (energetic particles sweeping out from the sun in all directions) will carry the wastes outward into the vastness of the space beyond the asteroid belt.

Where would we get the materials out of which to build power stations, factories, and, for that matter, observatories, laboratories, and large settlements in which to house the people working and living in space?

The major source, at least over the next century, would be the moon, which could supply concrete, glass, soil, every important metal, even oxygen. Of the key substances, only hydrogen, carbon, and nitrogen would have to be imported from Earth.

Within the next century, then, space could become the important cutting edge of human advance. It could be the industrial heart of humanity, while Earth's surface could be allowed to revert to a more wholesome ecological balance, with farmland, parks, and wilderness once more expanding. (It is even possible, by the way, that a good deal of human agriculture could be carried out in space, where sunshine is unlimited and the weather can be under perfect control.)

What's more, when it comes to the further exploration of the solar system and the emptiness beyond, it will probably be on the space settlers we will depend. They will be better accustomed to space and its rigors and psychologically more suited to long exploratory voyages. They will be the great mariners of space, extending still farther the range of humanity.

In short, we are now at the threshold of a vast new expansion of humanity in terms of range, knowledge, and ability—all of which can create an enormously different way of life (and, it is to be hoped, a better one) for our children and grandchildren.

What is needed now, more than anything else, is the vision that this might be so, and the willingness on the part of human beings of all nations to cooperate in the investment of time, effort, and money to make that vision a reality.

It would not take much time, effort, or money. Perhaps only one-tenth of the amount willingly invested by the nations of the world for the various competing military machines that can serve no purpose other than destruction, and the end of civilization.

# Part VII

# Personal

# 58

# I Am a Signpost

I am not a prophet by trade; I merely write science fiction.

However, if one writes science fiction and, in order to do so, thinks about the future in detail, one may become a prophet by inadvertence, and no one did it more inadvertently than I.

In May 1939, when I was but nineteen years old, I wrote a story I called "Robbie." It was about a robot of the year 1998. It was metallic. It looked vaguely human. It could not talk. As a casual part of the story, I also described a "talking robot" which, in order to gain the capacity for speech, was too large to be mobile. I described it as "an unwieldly, totally immobile mass of wires and coils spreading over twenty-five square yards."

It was clearly an electronic computer, but I had not bothered to foresee miniaturization, at least not by 1998. It was, I suppose a mass of bulky vacuum tubes, for the transistor had not yet been invented and I was not bright enough to predict it.

Yet I dimly recognized that there would have to be miniaturization of some sort, for mobile robots would have to have artificial brains of considerable complexity, capable of fitting inside a container not much larger than the human skull. (I never thought to call these brains "computers" till the real thing came along.)

I spoke, therefore, of "positronic brains" mentioning them first in a story called "Reason," which I wrote in November 1940. I didn't go into the exact engineering details, you understand, but the impression I left was that streams of positrons, constantly created and as constantly annihilated, had just enough time between appearance and disappearance to create delicate traceries that corresponded to whatever it was in human brains that marked out the electropotentials of thought.

Not bad, but why positrons? To be sure, they were only discovered six years before I wrote the story, so they sounded peculiarly science-fictional — but, heck, we do it all now with ordinary electrons. If I hadn't jumped far enough with my vacuum tubes, I jumped too far with my positrons.

I came closer to target when I realized that, if we were to develop computerized robots, we would have to develop the art and science of dealing with them. I abandoned the time-honored lone inventor who created his robot in the dank cellars of a mysterious castle. I brought into existence a research establishment that designed positronic brains (computers, that is) and, in a story named "Runaround," written in October 1941, I named the science "robotics."

Apparently, I was the first person in history to use that word, though I was not aware of that. I thought the word existed. It exists *now,* of course, and my often-repeated "Three Laws of Robotics," first explicitly stated in "Runaround," may well have helped bring the word into actual use.

World War II came and went and the real-life electronic computer came into being. I advanced my robots a notch to stay ahead. Once I heard of Univac, I invented the much more complex and intricate "Multivac," which, in story after story, dealt with more and more elaborate aspects of society.

It was mentioned first in my story "Franchise," written in December 1954, and there I described it as "half a mile long and three stories high." I said that "fifty technicians walked the corridors within its structure."

I was back to nonminiaturization — uncounted cubic yards of vacuum tubes.

Once again, however, the exigencies of my plots drove me to miniaturization before the fact. As early as 1950, I used pocket calculators in my stories and described them fairly accurately. At least, I described their *outside* appearance fairly accurately.

But then came my story "The Last Question," written in June 1956. In 5,700 words it outlined a trillion years of human history and it began with Multivac. By now I knew it was made up of transistors and not vacuum tubes, but it was still an enormous structure.

In the second scene of the story, however, I introduced "Microvac." It was every bit as complex as the largest Multivac but it was small enough to fit into an ordinary spaceship. "In place of transistors," I explained, "had come molecular valves."

I had finally moved beyond what human beings were to accomplish by 1981. We have the microchip, but even its components are not yet down to molecular size.

In the next stage "sub-mesons took the place of the old, clumsy molecular valves," and, before you ask — no, I don't know what sub-mesons are.

Computers, as they became still more elaborate in the later stages of "The Last Question," became utterly indescribable, for they moved into hyperspace. They became equally reachable from every point in space and yet truly reachable from nowhere. They were no longer under human control, for each computer designed its own far greater successor until, in the end, the computer had become—quite literally—God.

But what could I do *after* "The Last Question"? Having deified the computer, where were there new worlds to conquer? Well, in March 1975, I wrote "The Bicentennial Man," which was three times as long as "The Last Question," and covered a mere two centuries.

In "The Bicentennial Man," I painstakingly described the steps by which a computer became, not God, but something that touches us much more closely—a human being. The computer, having begun as the brain inside a robot, gained, little by little, a completely human appearance and a completely human behavior. Its final achievement was that of gaining mortality, so that it could prove its final humanity in the only way possible—by dying of old age.

It took me thirty-six years; and, in some fifty stories, ranging in length from short-shorts to novels, I think I must have touched, in one way or another, on every aspect of computers and computerization. And (mark this!) I did it without ever knowing anything at all about computers in any real sense.

To this day, I don't.

I am totally inept with machinery and, when someone asks me if I work with any computers myself, I shudder and say, "I am a signpost, sir. I point the way. I don't go there."

As it happens, almost every writer of my acquaintance is using a word-processor, or is getting one, while I cling (more or less in terror) to my electric typewriter.

Came the day when *Byte Magazine* asked me to write of my experiences with a word-processor, I bit my knuckle shyly and admitted I didn't have one.

The magazine frowned at me in corporate majesty and saw to it that one was installed in my apartment. It is there now. It frightens me. I toss about at night thinking of it.

Included are an encyclopedic set of directions that I am desperately reading. I have cassettes to which I am desperately listening. Soon I will be able to compose something on it and have it printed out, if I can control my trembling fits sufficiently.

And, once I do, I am told, I will *really* be able to put on some speed. On my typewriter I turn out books at the contemptible rate of one a month.

# The Word-Processor and I

In the previous chapter, "I Am a Signpost," I mentioned that a word-processor had entered my life and described the manner in which I faced it—head high, eyes flashing, fists clenched, and brain paralyzed with fear.

Let me give you the details. When *Byte* let it be known that, in their opinion, a word-processor would look good in Isaac Asimov's office, Ed Juge of Radio Shack, down in Fort Worth, Texas, thought, in his warm and loving heart, that that might be a good idea.* On May 6, 1981, therefore, a word-processor arrived. Or at least, two big boxes and a small one, each presumably filled with arcane incunabula, came.

I managed to hoist them from the lobby of the apartment house, where the delivery men left them, up to the thirty-third-floor apartment where I lived. Fortunately, that was not as difficult as it sounds, since I used the elevator.

I then placed the boxes in my office and practiced walking around them until I got the route memorized. To make sure, I practiced it in the dark, then with my eyes closed, then in the dark *and* with my eyes closed.

In a few hours, I was able to walk through my office without ever making contact with the boxes, or even looking in their direction. In this way, I was able to pretend that they didn't exist. Unfortunately, part of my library shelves were blocked by them, but I decided not to use those shelves. If I needed data contained in the books there, I could always make it up.

---

*Only today, I asked Ed (whom I have come to know and revere) whether he, down there in his Fort Worth office, ever recognized the existence of Dallas. "Do I ever recognize the existence of *what*?" he asked.

This worked fine and my heartbeat had come down to pretty nearly normal, when, on May 12, Ron Schwartz of Radio Shack arrived with the intention of emptying those boxes. With the help of my dear wife, Janet, he set up a "computer corner" in our living room. Within it, the word-processor was unboxed, hooked together, and plugged in. I did my bit, to be sure. I kept saying, "I don't think we have any space for a word-processor anywhere," but no one listened to me.

In no time at all, there it was—a Radio Shack TRS-80 Model II Micro-Computer, along with a Daisy-Wheel Printer and a Scripsit program. A bunch of floppy discs, some ribbons, and various other pieces of formidable paraphernalia were also included.

Ron then proceeded to show me how it worked. To me it seemed like a tremendously complex machine with a console reminiscent of that on a Boeing 707; but Ron, unconcerned, approached it in the most casual possible manner. He flicked the keys, and had me do the same, so that different things happened on the screen. Words and sentences appeared, and parts were then erased, substituted, transferred, inserted, started, stopped. Ron paused only to stifle a yawn or two.

"You see how simple it is," he said. "If you have any difficulties, here are two instruction booklets." (With an effort, he hoisted out two volumes, each the size of a Manhattan telephone directory.) "This one," he said, "comes with a series of cassettes so you can hear a nice friendly voice tell you everything there is to know. —And," he went on, "If reading and hearing the whole thing in detail isn't enough, just phone me. I'll be glad to repeat everything."

He left, and I spent the evening staring at the word-processor. Staring, it turned out, was not enough. No matter how hard I stared, and with what intensity I *thought* at it, it did nothing. With a sinking heart, I realized that there was no way out. I was going to have to fiddle with the keys.

In my diary that night, I wrote: "I'll *never* learn how to use it."

Nevertheless, the indomitable spirit of the Asimovs shone through. Painstakingly, I read the instructions, doing my best not to move my lips as I did so. I listened to several of the cassettes, trying to nod intelligently at odd moments.

It didn't help. On May 27, I wrote in my diary: "Very depressed because of the word-processor. Couldn't get it to double-space."

Hah! I also couldn't get it to make tables, or to reduce the margins to the insignificance I find comfortable. Nor, when I attempted to print one of my concoctions, could I get it to pause between pages, even though it promised it would. It would wait for me to prepare a page, with carbons; and, just before I could insert it, the thing would merrily begin imprinting the bare platen with my deathless prose.

On June 4, two young men from Radio Shack arrived, unfailingly pleasant, unfailingly helpful, unfailingly polite, and I wept on their shoulders in turn. Under their ministrations, the word-processor behaved like a purring little pussy-cat, but my heart told me they would not have been gone for as much as one and a half seconds before pussy-cat would turn back into a Bengal tiger. My heart was not wrong.

I had, by now, developed the habit of flinching when I passed the computer corner, and throwing up my arm as though to ward off an attack. Occasionally, I would open one of the instruction books and read the cheerful instructions at random, but it all echoed meaninglessly in the vacant cavity I austerely refer to as my brain.

On June 12, the word-processor had been sitting in my living room for a full month, and it had so far won every battle.

But was I downhearted? Did I feel beaten?

You bet I was, and you can also bet I did!

On June 14, I decided to make one last try before asking Radio Shack to remove the thing and take its beak from out my heart. I was going to attempt to write a short article on the word-processor. Actually, I had already written it in first draft on my trusty Selectric III typewriter, but it was my intention to transpose it to the screen, correct it, and then print it.

I sat down and started the machine—and, suddenly, with no warning whatever, everything worked. It rubbed its head against my leg and purred.

I will never know what happened. The day before I had been as innocent of the ability to run the machine as I had been while it had still been in its original box. A night had passed—an ordinary night—but during it something in my brain must finally have rearranged itself. Now, there I was, running the machine like an old hand. In making my corrections, I could even use my right hand on both the "repeat" and the cursor arrows, without looking, and that little blinking devil jumped through every hoop in sight.

On June 17, I took the big step. I had a massive manuscript of a book in first draft. I put the entire first chapter onto the screen and then printed it.

I whistled while I worked.

Janet (my dear wife) came by to watch, and stood there transfixed. I waved my hand airily at her. "Nothing to it," I said. "All it takes is grit, determination, a sense of buoyant optimism, and good old Yankee know-how."

I've been working at it steadily ever since (with some minor problems I'll tell you about another time). In a matter of days after the transformation, in fact, I called up Radio Shack and told them to send someone over to set up the special tables that had arrived after the word-processor itself had. "Put it all up," I said, "because I have decided I will keep the machine. I may even pay for it," I added, with devil-may-care insouciance.

Scott Stoegbauer of Radio Shack arrived on July 8 and did the job.

I said to him, "Very friendly machine you have here. Reliable. Easy to handle. Makes no trouble. All you need is raw courage and the kind of self-confidence that can surmount all hurdles."

"You are a great man, Dr. Asimov," said that very perceptive young fellow.

---

## 60

# A Question of Speed

In the previous chapter, "The Word-Processor and I," I told the saga of my conquest of a new technique through sheer grit and intelligence. And now that my beloved Radio Shack TRS-80 Model II Micro-Computer and its Scripsit™ Program are in place and working, the question is: How has this improved my writing?

All the writers I know who are using word-processors are unanimous on one point. "Wow!" they say, "How it has increased my speed! Revisions are a snap; no more endless reams of scrap paper, crumpled, torn, and piled up around my typewriter. Zip, zip, and an article it used to take me a month to do now takes me a bare week."

Then they clap me on the back and say, "You'll see, Asimov. Everything will go much faster for you now."

And all I can do is sit in the corner and brood, because things *can't* go any faster with me. Let me list my problems:

1. In the first place, I type quickly—90 words a minute, when I am happy, carefree and in a good mood. And that's my typing rate when I am composing, too, because I don't believe in fancy stuff. In my writing, there is no poetry, no complexity, no literary frills. Therefore, I need only barrel along, saying whatever comes to mind, and waving cheerfully at people who happen to pass my typewriter.

2. What revisions? I change words here and there, insert or remove commas more or less at random, and occasionally cross out a sentence or insert a clause, but I would say that 95 percent of what I write in the first draft stays in the second (and final) draft. When an article requires, let us say, ten pages, I end up with ten uncrumpled pages, and there is nothing piled up around my wastebasket, or in it either.

Conclusion: I *can't* speed up my writing.

Let me be specific. In the last 138 months, I have published 141 books, so that I am a one-man book-of-the-month club. How far can I speed up beyond that? In fact, who in the world would want me to?

What it amounts to is that the bottleneck is my poor brain.

For instance, I've been commissioned to write a mystery a month for a magazine called *Gallery*. The typing isn't hard. They're short mysteries and two or three hours of typing each month does the job. I've done twenty-three of them so far and, as far as the mechanics of typing are concerned, I can keep it up until old rocking chair gets me.

But how about thinking up a plot? I have to spend a lot of shower-time, shaving-time, taxi-time, falling-asleep time thinking them up and I honestly don't know how long I can maintain the pace. And how, for goodness sake, will a word-processor help with that?

Well, it won't, and it doesn't, but I am glad I have it anyway and I will tell you why.

First, the typing is less trouble. The touch is lighter and easier than even the best electric typewriter and, to my astonishment, I find I can actually move along even faster by a little bit (I'm afraid to time myself—I don't want to know). Furthermore, it is almost noiseless and, since my computer-corner is in my living-room, it is important that I not disturb my dear wife when I work. In fact, I can listen to television programs if I wish (and if I take quick peeks over my left shoulder, I can even see bits).

Second, revision *is* fun. My advisers are right in that respect. It is so easy to insert commas, for instance, that I have taken to inserting them every fifth word or so, whether I need them or not. Then, too, it is such a trifle to change word-order that I never have occasion to grow exasperated at the small infelicities I make in first draft. This results in a greater affection for the articles I write and makes it easier to treat any rejection with an appropriate contempt and scorn.

Third, speed is *not* the issue. Who cares about speed? Who needs speed? I've got plenty of speed, and all I ask is that the word-processor not slow me up—and it doesn't.

There's something more important than speed!

I have said that I type 90 words a minute, but I didn't say I type at that speed without errors. There are plenty of errors, and I learned not to be bothered by them. After all, I started my writing for the pulp-fiction market where payment had to be placed under a strong magnifying glass to be seen. One either typed quickly, without regard to errors, and collected many of these microscopic payments—or one stopped to correct errors, and starved.

The result is that any page I type may smoke slightly from the heat generated by the speed, but it is also garbled. The word "the" is spelled in various fashions—"eht," "eth," "teh," "th e" and so on. These are distributed, with fine impartiality, randomly over the page. I make few distinctions between "seep," "seen," and "seem" (or "esem" for that matter), and I am quite apt to mention "the button of the flask" when I am referring to that part of the flask at the opposite end from the top.

Naturally, I go over the final copy and correct all the typos I see. This has the effect of littering each page with pen-and-ink corrections. Even though my penmanship is reasonably neat and clear, the aesthetics of the page suffers as a result.

Then, too, such is my lovable slapdash nature that I suspect I am the third worst proofreader in North America. The amount of staring I usually allot to a page is just sufficient to enable me to find just about half the typos that are present. I can't stare any longer than that; there's always another article or story or book waiting to be written.

The result is that I get editorial calls as to the meaning of "snall paint" and I receive unsympathetic comments when I point out that I clearly meant "small print".

How different now! Staring at a page of type on a television screen, I find myself looking eagerly for typos so that I can have the fun of changing them. Bang goes the "F1" and the "u" and the "F2" and "cold" suddenly becomes "could" and no sign exists that it was ever anything else. I send the cursor flying, up and down, left and right, and all the "cart"s become "cat"s, and all the "hate"s become "heat"s (or vice versa). What's more, commas go zooming in by the thousands and interrogative sentences which, in the old days, had question marks attached at the rate of one in three, now have one inserted with loving care *every time*.

So I end up with letter-perfect copy and no one can tell it wasn't letter-perfect all the time. Then I have it printed — br-r-rp, br-r-rp, br-r-rp — and as each perfect page is formed, my heart swells with pride. Words cannot describe the disgust I feel when, on rare occasions, I let a typo slip by me and find that I have spelled "the" as "t4he." I moan pitifully as I cross out the 4 and make an ink-mark on the virginal page. (Of course I can revise the page and print up another copy, but I have not yet brought myself to do that.)

So it's not a question of speed after all, but of perfection. And I hope the copy-editors appreciate the new me, that's all.

# 61

# A Question of Spelling

I received a letter today from the "Reading Reform Foundation," which tells me that "23 million (American) adults are functionally illiterate, unable to read an advertisement, a job application, directions on a medicine bottle." They say "30 percent of all schoolchildren have serious reading difficulties."

I rather believe this, judging from my own limited experience with people. But *why* is this?

Can it be that part of the reason is the matter of English spelling? The letter tells me that "87 percent of English words are phonetic; each of them follows predictable rules for reading and spelling." But that means that 13 percent of English words are not, and that includes many common words indeed.

With spelling erratic, many English words become ideograms that must be learned as a whole, with its parts giving no clue or, worse yet, false clues. If you don't know in advance and just judge by the letters, can you know that "through," "coo," "do," "true," "knew," and "queue" all rhyme? If you don't know in advance and just judge by the letters, can you know that "gnaw," "kneel," "mnemonic" and "note" all start with the same consonantal sound?

Why can't we say "throo," "koo," "doo," "troo," "nyoo," and "kyoo"? Why can't we say "naw," "neel," "nemonik" and "note"?

It looks funny? Sure it does, because you've memorized the "correct" way — but millions are helped on the road to illiteracy because the "correct" way makes no sense.

The plural of "man" is "men." Why not mans"? Because its childish? Exactly! To say "mans" is the first impulse of children when they learn

340

plurals — the sensible impulse. But if "men" makes sense, why don't you ask for "two cen of soup"? Why do you ask for "cans"?

The fact is that any attempt to regularize our spelling and grammar in order to make it easier to read and speak English seems invariably to fail, in part because many millions, having invested (and wasted) countless hours in learning the rules, don't want to have to relearn them.

I don't entirely blame them. I don't want to have to relearn them myself. I've got spelling and grammar down pat, and, if the rules are changed, I will forever be misspelling words and misusing them. And yet there could be advantages to subjecting ourselves to the trouble, advantages that might outweigh the objections.

First, it could make it easier for following generations to learn to read and write, and I have been told by many idealists (including my own parents) that people are willing to undergo hardships in order that life might be easier for their children and grandchildren.

Second, it would help build a nation of more widespread literacy; a more educated nation, therefore; a more technologically advanced nation, therefore; a more powerful and prosperous nation, therefore. I've been told by idealists that Americans are willing to undergo personal discomfort to achieve all these aims.

Third, in a world as small as ours is today, as bound together economically, with every nation affecting every other willy-nilly, both for good and evil, and with the survival of civilization resting on nations in cooperation rather than in conflict, it would surely be helpful if as many people as possible could understand one another's speech. How convenient it would be if some particular language were the second language, if not the first, of the world's educated people. English comes closest to fulfilling that ideal. It is spoken as a first or second language by more people than any language but Chinese; and, while Chinese is confined almost entirely to eastern and southeastern Asia, English is spread across the world.

Yet can English become a truly global language? To be sure, there are nationalistic barriers to that, and many millions would refuse to speak English out of what would be called "patriotism." But, even if the need for world communication were to come to seem paramount, there would remain the erratic spelling and grammar of the English language, which serve as a *needless* discouragement.

But consider. Is reform to be an eternal impossibility? Technological advance affects the most intimate aspects of our way of life. The telephone did a great deal to curtail the art of letter-writing, so that I make it my business to use a half-size sheet of paper for all correspondence. The typewriter demolished the art of clear handwriting by producing a form of writing clearer still.

Is there something, then, that might destroy the idiosyncrasies of our otherwise admirable language? Might it be the home computer? I rather think it could conceivably be.

I reason it out as follows. From the very invention of writing fifty-five centuries ago, right down to the latest electric typewriter, all "dictionaries" have had to be in the head of the user who wielded the stylus, pen, or keyboard. The user had to know how to spell words; and, having learned them, he had to hold on to them exactly as he had learned them.

But suppose that there are more and more home computers with software that includes more and more word-processing units and that these replace other techniques for writing books, articles, reports, correspondence, and so on. The word-processing unit can (and undoubtedly will) include a dictionary that will *not* be in the head of the user for the first time in history.

Suppose that, as the years pass, the worst offenses against common sense are removed from our spelling system—"nite" in place of "night," for instance. Even that will be difficult, of course; I type "night" without even thinking, whereas "nite" will be a constant effort. The dictionary, however, will be geared for "nite"; it will not contain "night." That means that every time I forget and type "night," the dictionary will pick it up and indicate it as an error. I will correct it without much trouble and the need for such correction will grow smaller and smaller.

Successive editions of word-processing dictionaries will be put out with more and more extensive reforms of a type agreed upon by some "Academy of Spelling Reform," and adults, gradually and relatively painlessly, will be weaned away from nonphonetic spelling and needlessly irregular grammar. Young people, who will learn with the help of home computers from the start, will automatically accept whatever stage of reform they are introduced to, and eventually phonetic spelling and regular grammar will become characteristic of our language and this will lead, as I have said before, to a more literate, educated, prosperous nation, and to an English that may very well become a world language that will facilitate international cooperation and make more possible the survival of civilization.

It sounds foolishly idealistic, I suppose, and I must admit that I have a tendency to look at the rosy side of technological progress. For instance, I think that the home-computer industry won't be putting out reformed "dictionaries" in response to an independent movement for spelling reform. I have no hope for such an independent movement being powerful enough to achieve anything. I think that the home-computer industry may actually lead the way itself—demand reform—and push the new dictionaries.

Why? Why ever should the home-computer industry want to put itself out on a limb like that?

Simple. I think it is inevitable that computers be designed to read the written word, and reproduce it; and even to hear the spoken word and put it into print or follow its orders. This can be done with the language as it is, but how much easier it would be if spelling is phonetic and grammar is regular. In short, we may not be sensible enough to reform our language for the sake of ourselves and our children, but we may be much more likely to do it to make sure that computers are cheaper and more easily maintained.

---------------62---------------

# My Father

My father's life made a sharp right-angled turn in January 1923. Everything he had been and had had before — vanished.

My father, Judah Asimov, was born in a "shtetl" named Petrovichi, in Russia, about 250 miles southwest of Moscow, on December 21, 1896. He was Jewish, but the Tsarist oppression and its endemic anti-Semitism made itself felt there only in its negative aspects. There were places that Jews couldn't live, things they couldn't do, professions they couldn't enter. Jews in Petrovichi, however, were accustomed to such things and accepted the limits as a fact of life.

There were no active expressions of the anti-Semitism, however; at least, not in Petrovichi. There were no pogroms; there was no violence. My father played peacefully with Gentile boys.

What's more, both my father and mother were among the wealthier townspeople. My father's father owned a mill; my mother's father owned a general store. Both families were economically secure.

My father did not have an education in the ordinary European sense, of course. It was difficult indeed (though not impossible) for a Jew in Tsarist Russia to receive what we ordinarily think of as an education, let alone go to a university. However, my father had no ambitions in that direction. He went to a Hebrew school, where he was thoroughly grounded in biblical studies and Talmudic scholarship.

He learned to read and write not only Hebrew and Yiddish but Russian as well, and had an opportunity to become acquainted with Russian literature and to read enough secular Russian books to pick up a better education of the ordinary sort than the Gentile boys of the town had a chance to do. He also learned how to keep the books of his father's business, which meant a good acquaintance with the intricacies of arithmetic and calculation.

To be sure, World War I, the Russian Revolution, and the Civil War that followed were all unsettling, but again Petrovichi was fortunate. The German Army of Kaiser Wilhelm never reached quite as far east as Petrovichi, nor did the marauding White Russian bands reach quite as far north. The new Soviet government established itself quietly and firmly.

Petrovichi remained an island of relative calm and peace in the dreadful storm that was convulsing the land. In the wake of the fallen Tsarist autocracy, the new leaders proclaimed that all people would be equal, that education and cultural activities would be encouraged, that cooperative ventures would be welcomed, and so on, and my father took them at their word.

He therefore helped organize a library, and set up regular sessions in which he read aloud from Russian classics to those who were not able to do so for themselves. He participated in a drama group that produced classic works of Yiddish and Russian dramatists for Petrovichi and the surrounding town. He organized a cooperative that bought and distributed food in order to stave off hard times during a period of food shortage.

In all this, he was helped by the fact that the local Soviet functionary was a Gentile who, as a boy, had lived in Petrovichi and had played with my father. Indeed, my father had helped the boy with his homework.

And that was how things stood at the end of 1922.

What would have happened had things continued so? There were hard times ahead. Looming beyond the horizon were the Stalinist purges of the 1930s, the devastating calamity of the Nazi German invasion, the discomforts of the Soviet anti-Zionist stance.

My father might have survived it all. All my experience of him makes me certain that he would have acted always as a prudent man, carefully weighing his actions and never allowing himself to be misled by ambition into a dangerous exposure.

He was not of the stuff of storybook heroes and dreamed no impossible dreams. He judged the limits of the possible shrewdly and operated within those limits.

But within those limits, he might have led a happy, useful, and fruitful life—more educated than most of those about him, more intelligent, more driving, more quick to seize an opportunity or see a danger. He would have been an important and successful man in the limited sphere within which he would have chosen to operate, and he might have survived all dangers.

Except that, as 1922 came to an end, he was unexpectedly faced with a situation he had not foreseen, an opportunity he could not weigh, and he had to make a decision—

My mother had an older half-brother, Joseph Berman, who had emigrated to the United States some time before World War I and who lived in Brooklyn. With Russia racked by foreign invasions, revolutions, and civil war, he finally wrote to Petrovichi to inquire as to the welfare of his little sister.

When my mother wrote to him and assured him she and hers were well, my Uncle Joe invited her to come, with her husband and children, to the United States. He himself would supply the necessary sponsorship and the guarantee that the immigrants would not become a public charge. The immigration laws of the United States were still sufficiently liberal to make that possible.

There was a family council on the nature of the action to be taken.

*Against* emigration was the fact that my parents were comfortable and well off (by the standards of the place and time) and were living where they had lived all their lives, with their family, with their friends, with their steady happiness. Against emigration also was the fact that before they could leave the country they would have to get the permission of the government to do so, and surely the Soviet officials would be inclined to view the mere desire as evidence of disloyalty and act accordingly.

*For* emigration was the legend of the "golden land" of America, where Jews were free and everyone was rich — something my father was surely too cautious to accept at face value.

What agonies and uncertainties my father and mother endured I do not know. What arguments they made use of pro and con, what advice they received, what fears they entertained, what hopes they experienced, I can't say.

But the decision was made at last, however, and it was to emigrate. It was the only wild decision my father ever made, I think. They would leave their land, their home, their families and perhaps never see any of these things or people again (and they never did) and venture instead into a *terra incognita,* a land completely unknown.

Having made the decision, my father carried it through with determination. He began with his boyhood friend, who was the local Commissar, up through a higher functionary in Gomel, and to Moscow itself.

During the month of January 1923, my father, with my mother and two children (my three-year-old self and my younger sister), made the trip overland to Danzig, by way of Riga. From there a coastal steamer took us to Liverpool and then the liner *Baltic* took us to Ellis Island, which we reached on February 3, 1923.

The consequences of the decision were drastic and in many ways for the worse. My father was no longer a man of substance and reputation, looked up to and well thought of. In New York, he was virtually penniless and a faceless member of an uncountable crowd.

He applied for citizenship at once and looked forward to becoming an American citizen in five years, with the full rights of citizenship despite his birth and religion—but that did not alter the practical truth that he was a "greenhorn" and suddenly, for the first time in his life, *uneducated.*

In Russia, he could speak both Yiddish and Russian, could converse with equal ease whether he faced a Jew or a Gentile, was as knowledgeable about either Jewish or Russian folkways and customs as any person in the town or in the surrounding region.

In the United States, he could not speak a word of the dominant language; he could not read the street signs; he did not know the ways of the people.

He could turn for help to his fellow Jews; but, if they had been in the United States long enough to be able to help, they could not hide their amusement at his ignorance, or their contempt for it, and he burned with embarrassment. He was a stranger in a strange land, a person of no consequence at all, and with nothing in the way of skills out of which he could make anything but the barest living. In a very important sense, he found himself illiterate and mute.

Yet he managed to find work; and in 1926, three years after he had come to the United States, he had managed somehow to accumulate the money to buy a small mom-and-pop candy-store.

In that, and in other candy-stores over the next thirty years, my father and mother (and the children) worked 16 hours a day 7 days a week and counted out the profits by the pennies. My father managed to carry us through the Great Depression without missing a meal or being forced to turn to charity. He managed to send me and my younger American-born brother to college.

Eventually, when his children were grown and independent, my father sold the candy-store and took a part-time job instead. (By part-time, he meant 40 hours a week.)

He refused help at all times and would not take the money from either my brother or myself when we were prosperous and could well afford to let him have all he wanted. In the last year of his life, he retired to Florida on his own money, and there he died of natural causes on August 4, 1969, leaving my mother enough money to support her for the remaining four years of her life.

At no time did I ever hear him complain of his decision. He had not fled oppression or grinding poverty. He had come, I believe, because it was his considered intention to accept a lowered status for himself in order that his children end up with a heightened status. If that was his intention, it was fulfilled and—I am thankful to be able to say—he lived long enough to see it fulfilled. In particular, he lived to see his older son (myself) a university professor and the author of a hundred books.

And it was not America alone that made it possible. It was my father as well. I like to think there is enough in me for me to have been successful under any reasonable conditions—but can I be sure? With another father, with other goals set for me, what might I have become?

Jews, generally (according to the stereotype), value learning and encourage their children to enter the intellectual professions. How true that stereotype is, or how general it is, I do not know. Certainly I have met many Jews who are neither as intelligent or as learned as many Gentiles I have met.

But the special case of my father is beyond doubt. Perhaps because he had himself passed from a state of being educated and learned to a state of being virtually illiterate, he valued education and learning all the more. He could not regain what he had once had, but he was determined that *I* was to have it.

He would not let me read the magazines he sold to others, because he felt they would muddy my thinking—but he let me read science-fiction magazines, because he respected the word "science" and felt they would lure me into becoming a scientist—and he was right.

He had no money to buy me the things I wanted, and I knew that, and I rarely asked him for anything—but, when I was overwhelmed with desire for something that spelled "learning," he managed to find a way to let me have it. When I was eleven, he bought me a copy of the *World Almanac* for my birthday, when that was what I wanted. I might have cried myself sick for a baseball but I wouldn't have gotten it.

When I was fifteen, he managed to scrape together the funds to buy me a used typewriter when I wanted it. If I had wanted a bicycle, I might as well have asked for the moon. —And as soon as I submitted a story to a magazine, when I was eighteen, *and before I had actually sold one,* my father, on his own initiative, managed to find the money to buy me a *new* typewriter. That was an act of faith that staggered me.

Long afterward, when I was turning out book after book in steady progression, and giving a copy of each to my father as a matter of course, he looked through one of my more difficult science popularizations and finally managed to ask me something that must have long puzzled him.

"Isaac," he said, hesitantly, "where did you learn all this?"

"From you, Pappa," I said.

"From *me*?" he said. "I don't know one word about these things."

"Pappa," I said, "you taught me to value learning. That's all that counts. All these things are just details."

# Acknowledgments

"The Army of the Night" appeared in the *New York Times Magazine,* June 14, 1981, under the title "The 'Threat' of Creationism." ©1981 by The New York Times.

"Creationism and the Schools" appeared in *Penthouse,* January 1982, under the title "The Dangerous Myth of Creationism." ©1981 by Penthouse International.

"The Reagan Doctrine" appeared in the *Lone Star Review,* May 1981.

"The Blind Who Would Lead" appeared in *Maclean's,* February 2, 1981.

"Creeping Censorship" appeared in the *Christian Science Monitor,* April 29, 1982. ©1982 by The Christian Science Monitor.

"Losing the Debate" appeared in *SciQuest,* March 1982. ©1982 by the American Chemical Society.

"The Harvest of Intelligence" appeared in *Family Weekly,* June 29, 1980, under the title "Should We Try to Create a Brave New World?" ©1980 by Family Weekly, Inc.

"That Old-Time Violence" appeared in *TV Guide,* June 14, 1975, under the title "Violence— as Human as Thumbs." ©1975 by Triangle Publications, Inc.

"Little Green Men or Not" appeared in *TV Guide,* December 14, 1974, under the title "UFOs". ©1974 by Triangle Publications, Inc.

"Don't You Believe?" appeared in *Isaac Asimov's Science Fiction Magazine,* April 14, 1982. ©1982 by Davis Publications, Inc.

"Open Mind?" appeared in *SciQuest,* May-June 1982. ©1982 by the American Chemical Society.

"The Role of the Heretic" appeared in *Scientists Confront Velikovsky,* ed. by D. Goldsmith (W. W. Norton, 1977). ©1977 by Cornell University.

"The Good Earth Is Dying" appeared in *Der Spiegel,* no. 21, 1971, under the title "Die Gute Erde Stirbt."

"The Price of Survival" appeared in *Genesis,* June 1975.

"Letter to a Newborn Child" appeared in *Unicef News,* 73 (1975): 1.

"Technophobia" appeared in *New Jersey Bell Journal,* Spring 1982. ©1982 by New Jersey Bell.

"What Have You Done for Us Lately?" appeared in *Mechanix Illustrated,* March 1981, under the title "Has Technology Done Anything for Us Lately?"

"Is It Wise to Contact Advanced Civilizations?" appeared in *Second Look,* November 1978. Copyright 1978 by Second Look.

"Pure and Impure" appeared in *Saturday Review,* June 9, 1979. ©1979 by Saturday Review Magazine Corp.

"Do We Regulate Science?" appeared in *Health,* vol. 13, no. 9, under the title "Government vs. Science". ©1981 by Health Magazine.

"For Public Understanding of Science" appeared in *SIPIscope,* January-February 1982, under the title "Informing the Public: Why Bother?—The Case For." Copyright 1981 by Scientists Institute for Public Opinion.

"Science Corps" appeared in *Science Digest,* February 1982, under the title "The Science Corps Wants You." ©1982 by the Hearst Corporation.

"Science and Beauty" appeared in the *Washington Post,* August 12, 1979 under the title "Science and the Sense of Wonder." ©1979 by The Washington Post, Inc.

"Art and Science" appeared in *Design,* March-April 1978. ©1978 by the Saturday Evening Post Company.

"The Fascination of Science" appeared in the *Saturday Review,* August 1980, under the title "What's Fueling the Popular Science Explosion?" ©1980 by Saturday Review Magazine Corp.

"Sherlock Holmes as Chemist" appeared in *Science Digest,* August 1980. ©1980 by the Hearst Corporation.

349

"The Sky of the Satellites" appeared in *Science Digest,* October 1980, under the title "View from the Galilean Satellites." ©1980 by the Hearst Corporation.

"Neutron Stars" appeared in the *London Telegraph,* May 1, 1977.

"Black Holes" appeared in the *London Telegraph* sometime in 1979.

"Faster than Light" appeared in *Starlog,* December 1977.

"Hyperspace" appeared in *P.M.* (German-language), January 1981.

"Beyond the Universe" appeared in *Science Digest,* June 1980, under the title "What Lies Beyond the Universe?" ©1980 by the Hearst Corporation.

"Life on Earth" appeared in *Mobil Viewer's Guide "Life on Earth,"* 1982, under the title "Isaac Asimov Looks at 'Life on Earth.' "

"The Future of Collecting" appeared in *The Franklin Mint Almanac,* May 1980. Copyright 1979 by the Franklin Mint.

"The Computerized World" was syndicated by Field Enterprises, April 16, 1978, under the title "Wonderful World of Computers." ©1978 by Field Enterprises, Inc.

"The Individualism to Come" appeared in the *New York Times,* January 7, 1973. ©1973 by The New York Times.

"The Coming Age of Age" appeared in *Prism,* January 1975. ©1975 by the American Medical Association.

"The Decade of Decision" appeared in *Manchete* (Brazil, Portuguese-language), June 2, 1979.

"Do You Want to Be Cloned?" appeared in *Family Weekly,* March 4, 1979, under the title "Don't Be Afraid of Clones, They May Save Your Life." ©1979 by Family Weekly, Inc.

"The Hotel of the Future" appeared in *Signature,* November 1981, under the title "Checking in at Tomorrow's Hotel." ©1981 by the Diners Club, Inc.

"The Future of Plants" appeared in *Horticulture,* January 1980, under the title "The Future for Plants." ©1980 by the Massachusetts Horticultural Society.

"Bacterial Engineering" appeared in *Realites,* September-October 1979. ©1979 by Realites, USA Publications, Inc.

"Flying in Time to Come" was excerpted in *Flying,* September 1977, under the title "Flying in 2002." ©1977 by Ziff-Davis Publishing Company.

"The Ultimate in Communication" appeared in *Lifestyle,* December 1972, under the title "Person to Person." ©1972 by Lifestyle Publishing, Incorporated.

"His Own Particular Drummer" appeared in *Phi Delta Kappan,* September 1976.

"The Future of Exploration" appeared in *Explorer's Journal,* December 1980. ©1980 by the Explorers Club.

"Homo Obsoletus?" appeared in *Through the '80s,* ed. by Frank Feather (World Future Society, 1980). ©1980 by the World Future Society.

"Volatiles for the Life of Luna" appeared in *Future,* May 1975, under the title "Life on Luna—Why Not?"

"Touring the Moon" appeared in the *New York Times,* January 10, 1982, under the title "A Tourist's Guide to the Moon." ©1982 by The New York Times.

"Life on a Space Settlement" appeared in the *American Legion Magazine,* September 1980. Copyright 1980 by the American Legion.

"Payoff in Space" appeared in *GQ,* November 1980, under the title "Space, the Cutting Edge of Human Development." ©1980 by GQ Magazine Inc.

"I Am a Signpost" appeared in *Popular Computing,* November 1981. ©1981 by Popular Computing, Inc.

"The Word-Processor and I" appeared in *Popular Computing,* September 1982. ©1982 by Popular Computing, Inc.

"A Question of Speed" appeared in *Popular Computing,* June 1982. ©1982 by Popular Computing, Inc.

"A Question of Spelling" appeared in *Popular Computing,* July 1982. ©1982 by Popular Computing, Inc.